THE FRONTIERS COLLECTION

Series editors

THE FRONTIERS COLLECTION

The books in this collection are devoted to challenging and open problems at the forefront of modern science, including related philosophical debates. In contrast to typical research monographs, however, they strive to present their topics in a manner accessible also to scientifically literate non-specialists wishing to gain insight into the deeper implications and fascinating questions involved. Taken as a whole, the series reflects the need for a fundamental and interdisciplinary approach to modern science. Furthermore, it is intended to encourage active scientists in all areas to ponder over important and perhaps controversial issues beyond their own speciality. Extending from quantum physics and relativity to entropy, consciousness and complex systems—the Frontiers Collection will inspire readers to push back the frontiers of their own knowledge.

More information about this series at http://www.springer.com/series/5342

For a full list of published titles, please see back of book or springer.com/series/5342

Anthony Aguirre · Brendan Foster
Zeeya Merali
Editors

Trick or Truth?

The Mysterious Connection Between Physics and Mathematics

Springer

Editors
Anthony Aguirre
Department of Physics
University of California
Santa Cruz, CA
USA

Zeeya Merali
Foundational Questions Institute
Decatur, GA
USA

Brendan Foster
Foundational Questions Institute
Decatur, GA
USA

ISSN 1612-3018 ISSN 2197-6619 (electronic)
THE FRONTIERS COLLECTION
ISBN 978-3-319-27494-2 ISBN 978-3-319-27495-9 (eBook)
DOI 10.1007/978-3-319-27495-9

Library of Congress Control Number: 2015958338

Printed on acid-free paper

This Springer imprint is published by SpringerNature
The registered company is Springer International Publishing AG Switzerland

Preface

This book is a collaborative project between Springer and The Foundational Questions Institute (FQXi). In keeping with both the tradition of Springer's Frontiers Collection and the mission of FQXi, it provides stimulating insights into a frontier area of science, while remaining accessible enough to benefit a non-specialist audience.

FQXi is an independent, nonprofit organization that was founded in 2006. It aims to catalyze, support, and disseminate research on questions at the foundations of physics and cosmology.

The central aim of FQXi is to fund and inspire research and innovation that is integral to a deep understanding of reality, but which may not be readily supported by conventional funding sources. Historically, physics and cosmology have offered a scientific framework for comprehending the core of reality. Many giants of modern science—such as Einstein, Bohr, Schrödinger, and Heisenberg—were also passionately concerned with, and inspired by, deep philosophical nuances of the novel notions of reality they were exploring. Yet, such questions are often overlooked by traditional funding agencies.

Often, grant-making and research organizations institutionalize a pragmatic approach, primarily funding incremental investigations that use known methods and familiar conceptual frameworks, rather than the uncertain and often interdisciplinary methods required to develop and comprehend prospective revolutions in physics and cosmology. As a result, even eminent scientists can struggle to secure funding for some of the questions they find most engaging, while younger thinkers find little support, freedom, or career possibilities unless they hew to such strictures.

FQXi views foundational questions not as pointless speculation or misguided effort, but as critical and essential inquiry of relevance to us all. The institute is dedicated to redressing these shortcomings by creating a vibrant, worldwide community of scientists, top thinkers, and outreach specialists who tackle deep questions in physics, cosmology, and related fields. FQXi is also committed to engaging with the public and communicating the implications of this foundational research for the growth of human understanding.

As part of this endeavor, FQXi organizes an annual essay contest, which is open to everyone, from professional researchers to members of the public. These contests are designed to focus minds and efforts on deep questions that could have a profound impact across multiple disciplines. The contest is judged by an expert panel and up to 20 prizes are awarded. Each year, the contest features well over a hundred entries, stimulating ongoing online discussion long after the close of the contest.

We are delighted to share this collection, inspired by the 2015 contest, "Trick or Truth: The Mysterious Connection Between Physics and Mathematics." In line with our desire to bring foundational questions to the widest possible audience, the entries, in their original form, were written in a style that was suitable for the general public. In this book, which is aimed at an interdisciplinary scientific audience, the authors have been invited to expand upon their original essays and include technical details and discussion that may enhance their essays for a more professional readership, while remaining accessible to non-specialists in their field.

FQXi would like to thank our contest partners: Nanotronics Imaging, The Peter and Patricia Gruber Foundation, The John Templeton Foundation, and Scientific American. The editors are indebted to FQXi's scientific director, Max Tegmark, and managing director, Kavita Rajanna, who were instrumental in the development of the contest. We are also grateful to Angela Lahee at Springer for her guidance and support in driving this project forward.

2015

Anthony Aguirre
Brendan Foster
Zeeya Merali

Contents

Introduction

Anthony Aguirre, Brendan Foster and Zeeya Merali

The miracle of the appropriateness of the language of mathematics for the formulation of the laws of physics is a wonderful gift which we neither understand nor deserve.

Eugene Wigner (1960)[1]

Modern mathematics is the formal study of structures that can be defined in a purely abstract way, without any human baggage. Think of mathematical symbols as mere labels without intrinsic meaning. It doesn't matter whether you write "two plus two equals four", "2 + 2 = 4" or "dos mas dos igual a cuatro". The notation used to denote the entities and the relations is irrelevant; the only properties of integers are those embodied by the relations between them. That is, we don't invent mathematical structures – we discover them, and invent only the notation for describing them.

Max Tegmark (2014)[2]

Theoretical physics has developed hand-in-hand with mathematics. It seems almost impossible to imagine describing the fundamental laws of reality without recourse to a mathematical framework; at the same time, questions in physics have inspired many

[1]Wigner, E. in "The Unreasonable Effectiveness of Mathematics in the Natural Sciences," *Communications in Pure and Applied Mathematics* (John Wiley & Sons: 1960).
[2]Tegmark, M. in *Our Mathematical Universe: My Quest for the Ultimate Nature of Reality* (Random House: 2014).

A. Aguirre (✉)
Department of Physics, University of California, Santa Cruz, CA, USA
e-mail: aguirre@scipp.ucsc.edu

B. Foster · Z. Merali
Foundational Questions Institute, Decatur, GA, USA
e-mail: foster@fqxi.org

Z. Merali
e-mail: merali@fqxi.org

A. Aguirre et al. (eds.), *Trick or Truth?*, The Frontiers Collection,
DOI 10.1007/978-3-319-27495-9_1

discoveries in mathematics. In the seventeenth century, for instance, Isaac Newton and Gottfried Wilhelm von Leibniz independently developed calculus, a technique that formed the bedrock of much of Newtonian mechanics and became an essential tool for physicists in the centuries that followed. Newton laid the groundwork for the development of modern theoretical physics as an essentially mathematical discipline. By contrast, mathematics appears to play far less of an integral role in the other sciences.

Connections between pure mathematics and the physical world have sometimes only become apparent long after the development of the mathematical techniques, making the link seem even stranger. In the 1930s, for example, physicist Eugene Wigner realised that the abstract mathematics of group and representation theory had direct relevance to particle physics. By uncovering symmetry principles that relate different particles, physicists have since been able to predict the existence of new particles, which were later discovered. In 1960, Wigner wrote an essay pondering such correspondences entitled, "The Unreasonable Effectiveness of Mathematics in the Natural Sciences".

This issue is also of particular interest to FQXi. One of the Institute's scientific directors, Max Tegmark, has attempted to explain this intimate relationship by positing that physical reality is not simply represented by mathematics, it *is* mathematics. This "mathematical universe hypothesis" has been expounded in Tegmark's book, *Our Mathematical Universe: My Quest for the Ultimate Nature of Reality*, published in 2014. However, his view remains controversial.

So, why does there seem to be a mysterious connection between physics and mathematics? This is the question that FQXi posed in our 2015 essay contest. We asked entrants to consider whether the apparently special relationship between the two disciplines is real or illusory: trick or truth? The contest drew over 200 entries, from thinkers based in 41 countries across 6 continents, both within and outside the academic system. A key aim of these essay contests is to stimulate discussion and the online forum for the contest has generated over 7,000 comments, to date. This volume comprises 19 of the winning essays, which have been expanded and modified from their original forms by the authors, in part, to address points raised by commentators in the forums.

In Chap. 2, our first-prize winner, Sylvia Wenmackers, argues that the effectiveness of mathematical models at describing the physical world may not be as unreasonable as it appears at first sight. We must remember, she says, that humans are "children of the cosmos" who have evolved as part of the universe that we seek to describe. When this is taken into account, she says, our ability to model the world is far less strange.

Our joint second place winners, Matthew Saul Leifer and Marc Séguin, both wrote essays that explicitly address Tegmark's assertion that physics is mathematics. In Chap. 3, Leifer claims that, on the contrary, mathematics is physics. He explains how this most abstract of disciplines still has its roots in observations of the physical world. In Chap. 4, Séguin takes Tegmark's argument to what he says is its logical conclusion. He states that the hypothesis implies that we must live in a

"Maxiverse"—a multitude of universes in which every possible observation happens somewhere—and he outlines its implications.

Tommaso Bolognesi also considered Tegmark's mathematical universe hypothesis in his essay. Bolognesi was awarded the prize for "most creative presentation" for a fictional work, reproduced in Chap. 5, in which a detective attempts to solve crimes by considering whether there is some truth in the idea that physics is mathematics. Chapter 6 provides another inventive take on the question: Ian Durham presents two characters from Lewis Carroll's *Alice's Adventures in Wonderland* discussing the possibility that there are two realities, representational and tangible, and how to reconcile them. Durham's essay won FQXi's "entertainment" prize.

The authors of Chaps. 7–9 each use concrete examples to explain why mathematics has been so successful at describing the physical world, despite being the product of human creativity. Kevin Knuth discusses the derivation of additivity, while Tim Maudlin uses a different mathematical language to explain geometrical structure. Ken Wharton uses the example of the flow of time to demonstrate why branches of mathematics often develop faster than the physical models to which they are applied.

A number of entrants addressed Wigner's original claim concerning the "unreasonable effectiveness" of mathematics head on. In Chap. 10, Derek Wise argues that Wigner was wrong to assume that mathematics developed entirely independently from the physics that it is used to describe. In Chap. 11, David Garfinkle makes a similar case, noting that new mathematics is built on older mathematics, just as new models of physics are modifications of older ones, and that those older branches of physics and mathematics may have been developed in tandem, explaining the seemingly surprising connections between more recent developments. Nicolas Fillion, in Chap. 12, attempts to demystify the relationship between physics and mathematics by considering how mathematical models are constructed as approximations of reality. In Chap. 13, Noson Yonofsky argues that the relationship is entirely reasonable if you compare the symmetry principles that underlie both physics and mathematics. While in Chap. 14, Sophía Magnúsdóttir takes on the role of a "pragmatic physicist" to make the case that scientific models do not have to be mathematical to be useful.

The puzzle of how humans can make sense of physics, at all, is tackled in Chaps. 15 and 16. In their essay, Alexey Burov and Lev Burov consider the fine-tuning of the fundamental laws of nature that give rise not only to life, but also make the universe understandable to people. The "out-of-the-box thinking" prize was awarded to Sara Imari Walker who argued that the comprehensibility of the universe is not so astonishing if we accept that the evolution of the structure of reality will favour the development of states that are connected to other existing states in the universe.

Some entrants delved more deeply into the relationship between mathematics and the human mind. In Chap. 17, Christine Cordula Dantas argues that mathematics is an essential feature of how conscious minds understand themselves. Anshu Gupta Mujumdar and Tejinder Singh were awarded the "creative thinking" prize for their essay, presented in Chap. 18, which invokes cognitive science to explain how the abstract features of mathematics have seeds not in some Platonic plane, but within the brain.

Finally, in Chaps. 19 and 20, Philip Gibbs (the winner of our "non-academic" prize) and Cristinel Stoica use examples from modern physics to review many of the questions surrounding the mysterious connection between mathematics and physics—showing why it is so hard to uncover whether that relationship is a trick or truth.

This compilation brings together the writings of professional researchers and non-academics. The contributors to this volume include those trained in mathematics, physics, astronomy, philosophy and computer science. The contest generated some of our most imaginatively structured essays, as entrants strived to solve a mystery about the nature of reality that will no doubt remain with us for centuries to come: Does physics wear mathematics like a costume, or is math a fundamental part of physical reality?

Children of the Cosmos

Presenting a Toy Model of Science with a Supporting Cast of Infinitesimals

Sylvia Wenmackers

> *[A]ll our science, measured against reality, is primitive and childlike – and yet it is the most precious thing we have.*
>
> Albert Einstein [1, p. 404]
>
> *[...] I seem to have been only like a boy playing on the sea-shore, and diverting myself in now and then finding a smoother pebble or a prettier shell than ordinary, whilst the great ocean of truth lay all undiscovered before me.*
>
> Isaac Newton [2, p. 54]

Abstract Mathematics may seem unreasonably effective in the natural sciences, in particular in physics. In this essay, I argue that this judgment can be attributed, at least in part, to selection effects. In support of this central claim, I offer four elements. The first element is that we are creatures that evolved within this Universe, and that our pattern finding abilities are selected by this very environment. The second element is that our mathematics—although not fully constrained by the natural world—is strongly inspired by our perception of it. Related to this, the third element finds fault with the usual assessment of the efficiency of mathematics: our focus on the rare successes leaves us blind to the ubiquitous failures (selection bias). The fourth element is that the act of applying mathematics provides many more degrees of freedom than those internal to mathematics. This final element will be illustrated by the usage of 'infinitesimals' in the context of mathematics and that of physics. In 1960, Wigner wrote an article on this topic [4] and many (but not all) later authors have echoed his assessment that the success of mathematics in physics is a mystery.

The above quote is attributed to Isaac Newton shortly before his death (so in 1727 our shortly before), from an anecdote in turn attributed to [Andrew Michael] Ramsey by J. Spence [2]. See also footnote 31 in [3].

S. Wenmackers (✉)
KU Leuven, Centre for Logic and Analytic Philosophy, Institute of Philosophy,
Kardinaal Mercierplein 2—Bus 3200, 3000 Leuven, Belgium
e-mail: sylvia.wenmackers@hiw.kuleuven.be

S. Wenmackers
University of Groningen, Faculty of Philosophy, Oude Boteringestraat 52,
9712 GL Groningen, The Netherlands

A. Aguirre et al. (eds.), *Trick or Truth?*, The Frontiers Collection,
DOI 10.1007/978-3-319-27495-9_2

At the end of this essay, I will revisit Wigner and three earlier replies that harmonize with my own view. I will also explore some of Einstein's ideas that are connected to this. But first, I briefly expose my views of science and mathematics, since these form the canvass of my central claim.

Toy Model of Science

Science can be viewed as a long-lasting and collective attempt at assembling an enormous jigsaw puzzle. The pieces of the puzzle consist of our experiences (in particular those that are intersubjectively verifiable) and our argumentations about them (often in the form of mathematical models and theories). The search for additional pieces is part of the game. Any piece that we add to the puzzle at one time may be removed later on. Nobody knows how many pieces there are, what the shape of the border looks like, or whether the pieces belong to the same puzzle at all. We assume optimistically that this is the case, indeed, and we attempt to connect all the pieces of the puzzle that have been placed on the table so far.[1]

So, like Einstein and (allegedly) Newton in the quotes appearing on the title page, I view science as a playful and limited activity, which is at the same time a highly valuable and unprecedented one. Scientific knowledge is fallible, but there is no better way to obtain knowledge. Hence, it seems wise to base our other epistemic endeavors (such as philosophy) on science—a position known as 'naturalism'. In addition, there is no more secure foundation for scientific knowledge beyond science itself. The epistemic position of 'coherentism' lends support to the positive and optimistic project of science. It has been phrased most evocatively by Otto Neurath [7, p. 206]:

> Like sailors we are, who must rebuild their ship upon the open sea, without ever being able to put it in a dockyard to dismantle it and to reconstruct it from the best materials.[2]

We are in the middle of something and we are not granted the luxury of a fresh start. Hence, we cannot analyze the apparent unreasonable effectiveness of mathematics in science from any better starting point either. Condemned we are to deciphering the issue from the incomplete picture that emerges from the scientific puzzle itself, while its pieces keep moving. A dizzying experience.

Since I mentioned "mathematical models and theories", I should also express my view on those.[3] To me, mathematics is a long-lasting and collective attempt at

[1]Making these connections involves developing narratives. Ultimately, science is about storytelling. "The anthropologists got it wrong when they named our species *Homo sapiens* ('wise man'). In any case it's an arrogant and bigheaded thing to say, wisdom being one of our least evident features. In reality, we are *Pan narrans*, the storytelling chimpanzee."—Ian Stewart, Jack Cohen, and Terry Pratchett (2002) [6, p. 32].

[2]This is my translation of the German quote [7, p. 206]: "*Wie Schiffer sind wir, die ihr Schiff auf offener See umbauen müssen, ohne es jemals in einem Dock zerlegen und aus besten Bestandteilen neu errichten zu können.*"

[3]In the current context, I differentiate little between 'models' and 'theories'. For a more detailed account of scientific models, see [5].

thinking systematically about hypothetical structures—or imaginary puzzles, if you like. (More on this in section "Mathematics as Constrained Imagination" below.)

Selection Effects Behind Perceived Effectiveness of Mathematics in Physics

The four elements brought to the fore in this section collectively support my deflationary conclusion, that the effectiveness of mathematics is neither very surprising nor unreasonable.

A Natural History of Mathematicians

This section addresses the two following questions. What enables us to do mathematics at all? And how is it that we cannot simply describe real-world phenomena with mathematics, but even predict later observations with it? I think that we throw dust in our own eyes if we do not take into account to which high degree we—as a biological species, including our cognitive abilities that allow us to develop mathematics—have been selected by this reality.

To address the matter of whether mathematical success in physics is trick or truth (or something else), and in the spirit of naturalism and coherentism (section "Toy Model of Science"), we need to connect different pieces of the scientific puzzle. In the ancient Greek era, the number of available pieces was substantially smaller than it is now. Plato was amongst the first to postulate parallel worlds: alongside our concrete world, populated by imperfect particulars, he postulated a world of universal Ideas or ideal Forms, amongst which the mathematical Ideas sat on their thrones of abstract existence.[4] In this view, our material world is merely an imperfect shadow of the word of perfect Forms. Our knowledge of mathematics is then attributed to our soul's memories from a happier time, at which it had not yet been incarcerated in a body and its vista had not yet been limited by our unreliable senses.

This grand vision of an abstract world beyond our own has crippled natural philosophy ever since. The time has come to lay this view to rest and to search for better answers, guided by science. Although large parts of the scientific puzzle remain missing in our time, I do think that we are in a better position than the ancient Athenian scholars to descry the contours of an answer to the questions posed at the beginning of this section.

Let us first take stock of what is on the table concerning the origin of our mathematical knowledge. Is mathematical knowledge innate, as Plato's view implied? According to current science, the matter is a bit more subtle: mathematical knowledge

---

[4] I will have more to say on the ancient Greek view on mathematics and science in section "A Speculative Question Concerning the Unthinkable".

is not innate (unfortunately, since otherwise we would not have such a hard time learning or teaching mathematics), but there are robust findings that very young children (as well as newborns of non-human animals, for that matter) possess numerical abilities [8, 9]. So, we have innate cognitive abilities, that allow us to learn how to count and—with further effort—to study and to develop more abstract forms of mathematics.

This raises the further question as to the origin of these abilities. To answer it, we rely on the coherent picture of science, which tells us this: if our senses and reasoning did not work at all, at least to an approximation sufficient for survival, our ancestors would not have survived long enough to raise offspring and we would not have come into being. Among the traits that have been selected, our ancestors passed on to us certain cognitive abilities (as well as associated vices: more on this below). On this view, we owe our innate numerical abilities to the biological evolution of our species and its predecessors.

Let me give a number of examples to illustrate how our proto-mathematical capacities might have been useful in earlier evolutionary stages of our species. Being able to estimate and to compare the number of fruits hanging from different trees contributes to efficient foraging patterns. So does the recognition of regional and seasonal[5] patterns in the fruition of plants and the migration of animals. And the ability to plan future actions (rather than only being able to react to immediate incentives) requires a crude form of extrapolation of past observations. These traits, which turned out to be advantageous during evolution, lie at the basis of our current power to think abstractly and to act with foresight.

Our current abilities are advanced, yet limited. Let us first assess our extrapolative capacities: we are far from perfect predictors of the future. Sometimes, we fail to take into account factors that are relevant, or we are faced with deterministic, yet intrinsically chaotic systems. Consider, for example, a solar eclipse. An impending occultation is predicted many years ahead. However, whether the weather will be such that we can view the phenomenon from a particular position on the Earth's surface, that is something we cannot predict reliably a week ahead. Let us then turn to the more basic cognitive faculty of recognizing patterns. We are prone to patternicity, which is a bias that makes us see patterns in accidental correlations [10]. This patternicity also explains why we like to play 'connect the dots' while looking at the night sky: our brains are wired to see patterns in the stars, even though the objects we thus group into constellations are typically not in each other's vicinity; the patterns are merely apparent from our earthbound position.

In our evolutionary past, appropriately identifying many patterns yielded a larger advantage than the disadvantage due to false positives. In the case of a tiger, it is clear that one false negative can be lethal. But increasing appropriate positives invariable comes at the cost of increasing false positives as well.[6]

As a species, we must make do without venom or an exoskeleton, alas, but we have higher cognitive abilities that allow us to plan our actions and to devise mathematics.

[5]Or 'spatiotemporal', if you like to talk like a physicist.

[6]The same trade-off occurs, for instance, in medical testing and law cases.

These are our key traits for survival (although past success does not guarantee our future-proofness). In sum, mathematics is a form of human reasoning—the most sophisticated of its kind. When this reasoning is combined with empirical facts, we should not be perplexed that—on occasions—this allows us to effectively describe and even predict features of the natural world. The fact that our reasoning can be applied successfully to this aim is precisely why the traits that enable us to achieve this were selected in our biological evolution.

Mathematics as Constrained Imagination

In my view, mathematics is about exploring hypothetical structures; some call it the science of patterns. Where do these structures or patterns come from? Well, they may be direct abstractions of objects or processes in reality, but they may also be inspired by reality in a more indirect fashion. For instance, we could start from an abstraction of an actual object or process, only to negate one or more of its properties—just think of mathematics' ongoing obsession with the infinite (literally the non-finite). Examples involving such an explicit negation clearly demonstrate that the goal of mathematics is not representation of the real world or advancing natural science. Nevertheless, this playful and free exercise in pure mathematics may—initially unintended and finally unexpected—turn out to be applicable to abstractions of objects and processes in reality (completely different from the one we started from). Stated in this way, the effectiveness of mathematics surely seems unreasonable. However, I argue that there are additional factors at play that can explain this success—making these unintentional applications of mathematics more likely after all.

Let us return to the toy metaphor, assuming, for definiteness, that the puzzle of natural science appears to be a planar one. Of course, this is no reason for mathematicians not to think up higher dimensional puzzles, since their activity is merely imaginative play, unhindered by any of the empirical jigsaw pieces. However, it is plausible that the initial inspiration for considering, say, toroidal or hypercubic puzzles has been prompted by difficulties with fitting the empirical pieces into a planar configuration.[7] In addition, and irrespective of its source, this merely mathematical construct may subsequently prompt speculations about the status of the scientific puzzle. Due to feedback processes like these, the imaginative play is not as unconstrained as we might have assumed at the outset. The hypothetical structures of mathematics are not concocted in a physical or conceptual vacuum. Even in pure mathematics, this physical selection bias acts very closely to the source of innovation and creativity.

In the previous section, I highlighted that humans, including mathematicians, have evolved in this Universe. Mathematics itself also evolves by considering variations on earlier ideas and selection: this is a form of cultural evolution which allows changes on a much shorter time scale than biological evolution does. Just like in biology, this

[7] In this example, considering the negation of the planar assumption—rather than any of the other background assumptions—is prompted by troubles in physics.

variation produces many unviable results. Evolution is squandermanious—quite the opposite of efficient. The selection process is mainly driven by cultural factors, which are internal to mathematics (favoring theories that exhibit epistemic virtues such as beauty and simplicity). But, as we saw in the previous paragraph, empirical factors come into play as well, mediated by external interactions with science. Although mathematics is often described as an a priori activity, unstained by any empirical input, this description itself involves an idealization. In reality, there is no a priori.

Mathematics Fails Science More Often Than Not

For each abstraction, many variations are possible, the majority of which are not applicable to our world in any way. The effectiveness perceived by Wigner [4] may be due to yet another form of selection bias: one that makes us prone to focus on the winners, not the bad shots. Moreover, even scientific applications of mathematics that are widely considered to be highly successful have a limited range of applicability and even within that range they have a limited accuracy.

Among the mathematics books in university libraries, many are filled with theories for which not a single real world application has been found.[8] We could measure the efficiency of mathematics for the natural science as follows: divide the number of pages that contain scientifically applicable results by the total number of pages produced in pure mathematics. My conjecture is that, on this definition, the efficiency is very low. In the previous section we saw that research, even in pure mathematics, is biased towards the themes of the natural sciences. If we take this into account, the effectiveness of mathematics in the natural sciences does come out as unreasonable—unreasonably low, that is.[9]

Maybe it was unfair to focus on pure mathematics in the proposed definition for efficiency? A large part of the current mathematical corpus deals with applied mathematics, from differential equations to bio-statistics. If we measure the efficiency by dividing the number of 'applicable pages' by the total number of pages produced in all branches of mathematics, we certainly get a much higher percentage. But, now, the effectiveness of mathematics in the natural sciences appears reasonable enough, since research and publications in applied mathematics are (rightfully) biased towards real world applicability.

At this point, you may object that Wigner made a categorical point that there is some part of mathematics at all that works well, even if this does not constitute all or most of mathematics. I am sympathetic to this objection (and the current point is the least important one in my argument), but then what is the contrasting case: that

[8]This is fine, of course, since this is not the goal of mathematics.

[9]Here, I recommend humming a Shania Twain song: "So, you're a rocket scientist. That don't impress me much." If you are too young to know this song, consult your inner teenager for the appropriate dose of underwhelmedness.

no mathematics would describe anything in the Universe? I offer some speculations about this in section "A Speculative Question Concerning the Unthinkable".

Abundant Degrees of Freedom in Applying Mathematics: The Case of Infinitesimals

I once attended a lecture in which the speaker claimed that "There is a matter of fact about how many people are in this room". Unbeknownst to anyone else in that room, I was pregnant at the time, and I was unsure whether an unborn child should be included in the number of people or not. To me, examples like this show that we can apply mathematically crisp concepts (such as the counting numbers) to the world, but only because other concepts (like person or atom) are sufficiently vague.

The natural sciences aim to formulate their theories in a mathematically precise way, so it seems fitting to call them the 'exact sciences'. However, the natural sciences also allow—and often require—deviations from full mathematical rigor. Many practices that are acceptable to physicists—such as order of magnitude calculations, estimations of errors, and loose talk involving infinitesimals—are frowned upon by mathematicians. Moreover, all our empirical methods have a limited range and sensitivity, so all experiments give rise to measurement errors. Viewed as such, one may deny that any empirical science can be fully exact. In particular, systematic discrepancies between our models and the actual world can remain hidden for a long time, provided that the effects are sufficiently small, compared to our current background theories and empirical techniques.

Einstein put it like this: "As far as the laws of mathematics refer to reality, they are not certain; and as far as they are certain, they do not refer to reality" [11, p. 28]. To illustrate this point, I will concentrate on the calculus—the mathematics of differential and integral equations—and consider the role of infinitesimals in mathematics as well as in physics.

In mathematics, infinitesimals played an important role during the development of the calculus, especially in the work of Leibniz [12], but also in that of Newton (where they figure as 'evanescent increments') [13]. The development of the infinitesimal calculus was motivated by physics: geometric problems in the context of optics, as well as dynamical problems involving rates of change. Berkeley [14] ridiculed these analysts as employing "ghosts of departed quantities". It has taken a long time to find a consistent definition of infinitesimals that holds up to the current standards of mathematical rigour, but meanwhile this has been achieved [15]. The contemporary definition of infinitesimals considers them in the context of an incomplete, ordered field of 'hyperreal' numbers, which is non-Archimedean: unlike the field of real numbers, it does contain non-zero, yet infinitely small numbers (infinitesimals).[10] The alternative calculus based on hyperreal numbers, called 'non-standard analysis'

[10]Here, I mean by infinitesimals numbers larger than zero, yet smaller than $1/n$ for any natural number n.

(NSA), is conceptually closer to Leibniz's original work (as compared to standard analysis).

While infinitesimals have long been banned from mathematics, they remained in fashion within the sciences, in particular in physics: not only in informal discourse, but also in didactics, explanations, and qualitative reasoning. It has been suggested that NSA can provide a post hoc justification for how infinitesimals are used in physics [16]. Indeed, NSA seems a very appealing framework for theoretical physics: it respects how physicists are already thinking of derivatives, differential equations, series expansions, and the like, and it is fully rigorous.[11]

Rephrasing old results in the language of NSA may yield new insights. For instance, NSA can be employed to make sense of classical limits in physics: classical mechanics can be modelled as quantum mechanics with an infinitesimal Planck constant [22]. Likewise, Newtonian mechanics can be modelled as a relativity theory with an infinite maximal speed, c (or infinitesimal $1/c$).

Infinitesimal numbers are indistinguishable from zero (within the real numbers), yet distinct from zero (as can be made explicit in the hyperreal numbers). This is suggestive of a physical interpretation of infinitesimals as 'currently unobservable quantities'. The ontological status of unobservables is an important issue in the realism–anti-realism debate [23]. Whereas constructive empiricists interpret 'observability' as 'detectability by the human, unaided senses' [24], realists regard 'observability' as a vague, context-dependent notion [25]. When an apparatus with better resolving power is developed, some quantities that used to be unobservably small become observable [26, 27]. This shift in the observable-unobservable distinction can be modelled by a form of NSA, called relative analysis, as a move to a finer context level [28]. Doing so requires the existing static theory to be extended by new principles that constrain the allowable dynamics [29].

The interpretation of (relative) infinitesimals as (currently) unobservable quantities is suggestive of why the calculus is so applicable to the natural sciences: it appears that infinitesimals provide scientists with the flexibility they need to fit mathematical theories to the empirically accessible world. To return to the jigsaw puzzle analogy of section "Toy Model of Science": we need some tolerance at the edges of the pieces. If the fit is too tight, it becomes impossible to connect them at all.

A Speculative Question Concerning the Unthinkable

Could our cosmos have been different—so different that a mathematical description of it would have been fundamentally impossible (irrespective of whether life could

[11] It has been shown that physical problems can be rephrased in terms of NSA [17], both in the context of classical physics (Lagrangian mechanics [18]) and of quantum mechanics (quantum field theory [19], spin models [20], relativistic quantum mechanics [21], and scattering [18]). Apart from formal aspects (mathematical rigour), such a translation also offers more substantial advantages, such as easier (shorter) proofs.

emerge in it)? Some readers may have the impression that I have merely explored issues in the vicinity of this mystery, without addressing it directly.

Before I indulge in this speculation, it may be worthwhile to remember that the very notion of a 'cosmos' emerged in ancient Greek philosophy, with the school of Pythagoras, where it referred to the order of the Universe (not the Universe itself). It is closely related to the search for *archai* or fundamental ordering principles. It is well known that the Pythagorians took the whole numbers and—by extension—mathematics as the ordering principle of the Universe. Their speculations about a mathematically harmonious music of the spheres resonated with Plato and Johannes Kepler (the great astronomer, but also the last great neoplatonist). Since these *archai* had to be understandable to humans, without divine intervention or mystical revelation, they had to be limited in number and sufficiently simple. So, the idea that the laws of nature have to be such that they can be printed on the front of a T-shirt, goes back to long before the invention of the T-shirt.[12] In this sense, the answer to the speculative question at the start of this section is 'no' and trivially so, for otherwise it would not be a cosmos. Yet, even if we understand 'our cosmos' as 'the Universe', there is a strong cultural bias to answer the speculative question in the negative.

In section "A Natural History of Mathematicians", I considered our proto-mathematical abilities as well as their limits. At least in some areas, our predictions do better than mere guesses. This strongly suggests that there are patterns in the world itself—maybe not the patterns that we ascribe to it, since these may fail, but patterns all the same. It is then often taken to be self-evident that these patterns must be mathematical, but to me this is a substantial additional assumption. On my view of mathematics, the further step amounts to claiming that nature itself is—at least in principle—understandable by humans. I think that all we understand about nature are our mathematical representations of it.[13] Ultimately, reality is not something to be understood, merely to be. (And, for us, to be part of.)

When we try to imagine a world that would defy our mathematical prowess, it is tempting to think of a world that is totally random. However, this attempt is futile. Pure randomness is a human idealization of maximally unpredictable outcomes (like a perfectly fair lottery [30]). Yet, random processes are very well-behaved: they consist of events that may be maximally unpredictable in isolation, but collectively they produce strong regularities. It is no longer a mystery to us how order emerges from chaos. In fact, we have entire fields of mathematics for that, called probability theory and statistics, which are closely related to branches of physics, such as statistical mechanics.

As a second attempt, we could propose a Daliesque world, in which elements combine in unprecedented ways and the logic seems to change midgame: rigid clocks become fluid, elephants get stilts, and tigers emerge from the mouths of fish shooting

---

[12]In case this remark made you wonder: the T-shirt was invented about a century ago.

[13]My view of mathematics might raise the question: "Why, then, should we expect that anything as human and abstract as mathematics applies to concrete reality?" I think this question is based on a false assumption, due to prolonged exposure to Platonism—remnants of which are abundant in our culture.

from a pomegranate. Yet, even such surrealistic tableaus have meta-regularities of their own. Many people are able to recognize a Dalí painting instantly as his work, even if they have not seen this particular painting before. Since we started from human works of art, unsurprisingly, the strategy fails to outpace our own constrained imagination.

At best, I can imagine a world in which processes cannot be summarized or approximated in a meaningful way. Our form of intelligence is aimed at finding the gist in information streams, so it would not help us in this world (in which it would not arise spontaneously by biological evolution either). In any case, what I can imagine about such a world remains very vague—insufficient for any mathematical description. Maybe there are better proposals out there?

Max Tegmark has put forward an evocative picture of the ultimate multiverse as consisting of all the orderings that are mathematically possible [31]. (See also Marc Séguin's contribution [32].) Surely, this constitutes a luscious multiplicity. From my view of mathematics as constrained imagination, however, the idea of a mathematical multiverse is still restricted by what is thinkable by us, humans. Aristotle described us as thinking animals, but for the current purpose 'mathematizing mammals' may fit even better. My diagnosis of the situation is that the speculative question asks us to boldly go even beyond Tegmark's multiverse and thus to exceed the limits of our cognitive kung fu: even with mathematics, we cannot think the unthinkable.

Reflections on Wigner and Einstein

In this section, I return to Wigner [4] and compare my own reflections to earlier replies by Hamming [33], Grattan-Guinness [34], and Abbott [35]. I end with a short reflection on some ideas of Einstein [11] that predate Wigner's article by four decades.

Wigner's Two Miracles

Wigner wrote about "two miracles": "the existence of the laws of nature" and "the human mind's capacity to divine them" [4, p. 7]. First and foremost, I hope that my essay helps to see that we need not presuppose the former to understand the latter: it is by assuming that the Universe forms a cosmos that we have started reading laws into it. Galileo Galilei later told us that those laws are mathematical ones. The very term "laws of nature" may be misleading and for this reason, I avoided it so far (except for the remark of fitting them on a T-shirt). The fact that our so-called laws can be expressed with the help of mathematics should be telling, since that is *our* science of patterns. When we open Galileo's proverbial book of nature, we find it filled with our own handwriting.

To illustrate the unreasonable effectiveness of mathematics, Wigner offered the following analogy for it: consider a man with many keys in front of many doors, who "always hits on the right key on the first or second try" [4, p. 2]. Lucky streaks like this may seem to require further explanation. However, if there are many people, each with many keys, it becomes likely that at least one of them will have an experience like Wigner's man—and no further explanation is needed (see also Hand [36]). There are indeed many people active in mathematics and science, and few of them succeed "on the first or second try"—or at all. In the essay, I argued that the perceived effectiveness of mathematics in physics can be diagnosed in terms of selection bias. The same applies to the metaphor Wigner presented.

Wigner found it hard to believe that the perfection of our reasoning power was brought about by Darwin's process of natural selection [4, p. 3]. Ironically, the selection bias that he may have fallen prey to is a good illustration of the *lack* of perfection of our reasoning powers. To be clear, I do not claim that Darwin's theory of biological evolution suffices to explain the success of (certain parts of) mathematics in physics. However, a similar combination of variation and selection is at work in the evolution of mathematics and science. See also Pólya, as cited by Grattan-Guinness, who has given an iterative or evolutionary description of the development of science [37, Vol. 2, p. 158].

Previous Replies to Wigner

In his description of mathematics, Wigner wrote about "defining concepts beyond those contained in the axioms" [4, p. 3]. Wigner did not, however, reflect on where those axioms come from in the first place. This has been criticized by Hamming [33], a mathematician who worked on the Manhattan Project and for Bell Labs. Axioms or postulates are not specified upfront; instead, mathematicians may try various postulates until theorems follow that harmonize with their initial vague ideas. Hamming cited the Pythagorean theorem and Cauchy's theorem as examples: if mathematicians would have started out with a system in which those crucial results would not hold, then—according to Hamming—they would have changed their postulates until they did. And, of course, the initial vague ideas are thoughts produced by beings entrenched in the physical world. This brings us to Putnam, who pointed out that mathematical knowledge resembles empirical knowledge in many respects: "the criterion of success in mathematics is the success of its ideas in practice" [38, p. 529].

Wigner did concede that not any mathematical concept will do for the formulation of laws of nature in physics [4, p. 7], but he claimed that "in many if not most cases" these concepts were developed "*independently* by the physicist and *recognized* then as having been conceived before by the mathematician" [4, p. 7] (my emphasis). I think this part is misleading: there is a lot of interaction between mathematics and physics and what is being 'recognized' is actually the finding of a new analogy. We may think that we are merely discovering a similarity, when we are really forging new connections, which may subtly alter both sides. This is a creative element of great importance within mathematics as well as in finding applications to other fields, in which our patternicity may be a virtue rather than a vice.

Both aspects have been illuminated by Grattan-Guinness, a historian of mathematics, who argued that "much mathematics has been motivated by interpretations in the sciences" [34, p. 7]. He stressed the importance of analogies within mathematics and between mathematics and natural science and he gave historical examples in which mathematics and physics take turns in reshaping earlier concepts. Moreover, he remarked that there are many analogies that can be tried (somewhat similar to the ideas in section "Abundant Degrees of Freedom in Applying Mathematics: The Case of Infinitesimals"), but that only the successful ones are taken into account when assessing the effectiveness of mathematics (postselection as in section "Mathematics as Constrained Imagination").

My essay mainly focused on elementary mathematics and simple models. Of course, there are very complicated mathematical theories in use in advanced physics. In relation to this, Grattan-Guinness observed that "by around 1900 linearisation had become something of a fixation" [34, p. 11], but he also discussed the subsequent "desimplification" or "putting back in the theory effects and factors that had been deliberately left out" [34, p. 13].

In light of my discussion of infinitesimals in this essay, it is curious to observe that Grattan-Guinness and Hamming referred to them too. Grattan-Guinness spoke approvingly of the Leibniz-Euler approach to the calculus because it "often has a better analogy content to the scientific context" [34, p. 15]. Hamming even mentioned NSA, but only as an example of the observation that "logicians can make postulates that put still further entities on the real line" [33, p. 85].

Recently, the reply by Hamming has been developed further by Abbott, a professor in electrical engineering [35]. Whereas Hamming described his recurrent feeling that "God made the universe out of complex numbers" [33, p. 85], Abbott described the complex numbers as "simply a convenience for describing rotations", concluding that "the ubiquity of complex numbers is not magical at all" [35, p. 2148].

More specifically, Abbott adds two points to Hamming's earlier observations. Abbott's first addition is that "all physical laws and mathematical expressions of those laws are [...] necessarily compressed due to the limitations of the human mind" [35, p. 2150]. He explains that the associated loss of information does not preclude usefulness "provided the effects we have neglected are small", which lends itself perfectly to a rephrasing in terms of 'my' relative infinitesimals. Abbott's second addition is that "the class of successful mathematical models is preselected", which he described as a "Darwinian selection process" [35, p. 2150]. Like I did here, Abbott warned his readers not to overstate the effectiveness of mathematics. Moreover, as an engineer, he is well aware that "when analytical methods become too complex, we simply resort to empirical models and simulations" [35, p. 2148].

The title of Abbot's piece, "The reasonable ineffectiveness of mathematics", and the general anti-Platonist stance agree with the views exposed in the current essay. In addition, Abbott tried to show that this debate is relevant, even for those who prefer to "shut up and calculate", because "there is greater freedom of thought, once we realize that mathematics is something we entirely invent as we go along" [35, p. 2152].

Einstein's Philosophy of Science

In 1921, so almost forty years before Wigner wrote his article, Einstein gave an address to the Prussian Academy of Sciences in Berlin titled "Geometrie und Erfahrung". The expanded and translated version contains the following passage (which includes the sentence already quoted in section "Abundant Degrees of Freedom in Applying Mathematics: The Case of Infinitesimals"):

> At this point an enigma presents itself which in all ages has agitated inquiring minds. How can it be that mathematics, being after all a product of human thought which is independent of experience, is so admirably appropriate to the objects of reality? Is human reason, then, without experience, merely by taking thought, able to fathom the properties of real things. In my opinion the answer to this question is, briefly, this: As far as the laws of mathematics refer to reality, they are not certain; and as far as they are certain, they do not refer to reality. [11, p. 28]

On the one hand, we read of "an enigma", which seems to set the stage for Wigner's later question. On the other hand, we are not dealing with a Platonist conception of mathematics here, since Einstein describes it as "a product of human thought". Yet, how are we to understand Einstein's addition that mathematics is "independent of experience"? This becomes clearer in the remainder of his text: geometry stems from empirical "earth-measuring", but modern axiomatic geometry, which allows us to consider multiple axiomatisations (including non-Euclidean ones), remains silent on whether any of these axiom schemes applies to reality [11, pp. 28–32]. Einstein refers approvingly to Schlick's view that axioms in the modern sense function as implicit definitions, which is in agreement with Hilbert's formalist position.[14]

Regarding his own answer to the question, Einstein further explains that axiomatic geometry can be supplemented with a proposition to relate mathematical concepts to objects of experience: "Geometry thus completed is evidently a natural science" [11, p. 32]. My essay agrees with Einstein's answer: the application of mathematics supplies additional degrees of freedom external to mathematics and we can never be sure that the match is perfect, since empirical precision is always limited. On other occasions, Einstein also pointed out that Kant's a priori would better be understood as 'conventional': a position close to that of Pierre Duhem (see for instance [39]).

In my late teens, I read a Dutch translation of essays by Einstein. The oldest essay in that collection stemmed from 1936 [40], so the text from which I quoted does not appear in that book. However, it is clear that similar ideas have influenced me in my formative years. It was a pleasure to reexamine some of them here. They contributed to my decision to become a physicist and a philosopher of physics, which in turn helped me to write this essay. Hence, the appearance of a text in this collection that is in agreement with some of Einstein's view on science may be a selection effect as well.

[14]For the influence of Moritz Schlick on Einstein's ideas, see [39].

Conclusion

In this essay, I have argued that:

- we are selected (A Natural History of Mathematicians);
- our mathematics is selected (Mathematics as Constrained Imagination);
- the application of mathematics has degrees of freedom beyond those internal to mathematics (Abundant Degrees of Freedom in Applying Mathematics: The Case of Infinitesimals);
- and, still, effective applications of mathematics remain the exception rather than the rule (Mathematics Fails Science More Often Than Not).

Too often, physicists have linearly approximated phenomena and considered themselves gods, only to discover much richer overtones of these phenomena later on. Let this be a lesson in modesty. While we are playing, things may appear to be very simple, but we should beware that (paraphrasing J.L. Austin[15]): it is not nature, it is scientists that are simple.

In the same spirit, I do not claim that my essay answers all questions concerning the relationship between mathematics and physics. Large gaps do remain between the pieces of this puzzle, but I found pleasure in arranging them in a way that suggests that further connections can be made. Feel free to pick up the pieces where I left them. Consider this essay your invitation to start playing.

Disclaimer
No parallel universes were postulated during the writing of this essay.

Acknowledgments Science is a multiplayer game. Therefore, I am grateful to Danny E.P. Vanpoucke for his feedback on an earlier version of this essay and to all participants in the discussion on the FQXi forum [42]. However, we do play with real money. This work was financially supported by a Veni-grant from the Dutch Research Organization (NWO project "Inexactness in the exact sciences" 639.031.244). I am grateful to FQXi for organizing the 2015 essay contest "Trick or Truth: the Mysterious Connection Between Physics and Mathematics", thereby giving me an incentive to write this piece. And, of course, I am very thankful that they awarded me the first prize for it.

References

1. A. Einstein, letter to Hans Muesham, July 9, 1951; Einstein archives 38–408; cited. In: Alice Calaprice (ed.) *The Ultimate Quotable Einstein*, Princeton, N.J: Princeton University Press (2011).
2. J. Spence, *Anecdotes, Observations, and Characters of Books and Men* (1820).
3. A.A. Martinez, *Science Secrets*, University of Pittsburgh Press (2011) 274–275.
4. E.P. Wigner "The unreasonable effectiveness of mathematics in natural sciences", *Communications on Pure and Applied Mathematics* **13** (1960) 1–14.

[15]Original quote by J.L. Austin, 1979 [41]: "It's not things, it's philosophers that are simple."

5. S. Wenmackers and D.E.P. Vanpoucke, "Models and simulations in material science: two cases without error bars", *Statistica Neerlandica* **66** (2012) 339–355.
6. I. Stewart, J. Cohen, and T. Pratchett, *The Science of Discworld II: The Globe*, Ebury Press (2002).
7. O. Neurath, "Protokollsätze", *Erkenntnis* **3** (1933) 204–214.
8. S. Dehaene, *The Number Sense; How the Mind Creates Mathematics*, Oxford University Press (1997).
9. R. Rugani, G. Vallortigara, K. Priftis, and L. Regolin, "Number-space mapping in the newborn chick resembles humans' mental number line", *Science* **347** (2015) 534–536.
10. M. Shermer, "Patternicity", *Scientific American* **299** (2008) 48.
11. A. Einstein, "Geometry and experience", translated by G. B. Jeffery and W. Perrett, in: *Sidelights on Relativity*, London, Methuen (1922) pp. 27–56.
12. G.W. Leibniz, "Nova Methodus pro Maximis et Minimis, Itemque Tangentibus, qua nec Fractas nec Irrationalales Quantitates Moratur, et Singular pro illi Calculi Genus", Acta Eruditorum (1684).
13. I. Newton, "Introductio ad Quadraturum Curvarum", chapter. In: *Opticks or, a Treatise of the Reflexions, Refractions, Inflexions and Colours of Light. Also Two Treatises of the Species and Magnitude of Curvilinear Figures*, London (1704).
14. G. Berkeley, *The Analyst; or a Discourse Addressed to an Infidel Mathematician*, printed for J. Tonson in the Strand, London (1734).
15. A. Robinson, *Non-Standard Analysis*, North-Holland, Amsterdam (1966).
16. S. Sanders, "More infinity for a better finitism", *Annals of Pure and Applied Logic* **161** (2010) 1525–1540.
17. S. Albeverio, J.E. Fenstad, R. Høegh-Krohn, and T. Lindstrøm (eds), *Nonstandard Methods in Stochastic Analysis and Mathematical Physics*, Academic Press, Orlando (1986).
18. F. Bagarello and S. Valenti, "Nonstandard analysis in classical physics and quantum formal scattering", *International Journal of Theoretical Physics* **27** (1988) 557–566.
19. P.J. Kelemen, "Quantum mechanics, quantum field theory and hyper-quantum mechanics", chapter. In: Victoria Symposium on Non-Standard Analysis, Lecture Notes in Mathematics, Vol. 369, Springer, Berlin (1974).
20. L.L. Helms and P.A. Loeb, "Application of nonstandard analysis to spin models", *Journal of Mathematical Analysis and Applications* **69** (1969) 341–352.
21. C.E. Francis, "Application of nonstandard analysis to relativistic quantum mechanics", Journal of Physics A 14 (1981) 2539–2251.
22. R.F. Werner and P.H. Wolff, "Classical mechanics as quantum mechanics with infinitesimal $\hbar$", *Physics Letters A* **202** (1995) 155–159.
23. A. Kukla, "Observation", chapter. In: S. Psillos and M. Curd (eds), The Routledge Companion to Philosophy of Science, Routledge, London (2010) 396–404.
24. B.C. van Fraassen, *The Scientific Image*, Oxford University Press (1980).
25. G. Maxwell, "The ontological status of theoretical entities", chapter. In: H. Freigl and G. Maxwell (eds), Scientific Explanation, Space, and Time, Minnesota Studies in the Philosophy of Science, Vol. 3, University of Minnesota Press, Mineapolis (1962) 3–15.
26. I. Douven, *In defence of scientific realism*, Ph.D. dissertation, University of Leuven (1996).
27. T.A.F. Kuipers, *From Instrumentalism to Constructive Realism*, Synthese Library: Studies in Epistemology, Logic, Methodology, and Philosophy of Science, Vol. 287, Kluwer, Dordrecht (2000).
28. K. Hrbacek, O. Lessman, and R. O'Donovan, "Analysis with ultrasmall numbers", *The American Mathematical Monthly* **117** (2010) 801–816.
29. S. Wenmackers, "Ultralarge lotteries: analyzing the Lottery Paradox using non-standard analysis", *Journal of Applied Logic* **11** (2013) 452–467.
30. S. Wenmackers and L. Horsten, "Fair infinite lotteries", Synthese 190 (2013) 37–61.
31. M. Tegmark, "The Mathematical Universe", *Foundations of Physics* **38** (2008) 101–150.
32. M. Séguin, "My God, It's Full of Clones: Living in a Mathematical Universe", Chapter 4 in current book.

33. R. W. Hamming, "The Unreasonable Effectiveness of Mathematics", *The American Mathematical Monthly* **87** (1980) 81–90.
34. I. Grattan-Guinness, "Solving Wigner's mystery: The reasonable (though perhaps limited) effectiveness of mathematics in the natural sciences", The Mathematical Intelligencer (2008) 7–17.
35. D. Abbott, "The Reasonable Ineffectiveness of Mathematics", *Proceedings of the IEEE* **101** (2013) 2147–2153.
36. D. Hand, *The Improbability Principle; Why Coincidences, Miracles, and Rare Events Happen Every Day*, Scientific American / Farrar, Straus and Giroux (2014).
37. G. Pólya, *Mathematics and Plausible Reasoning*, Princeton University Press (1954; 2nd edition 1968).
38. H. Putnam, "What is mathematical truth?", *Historia Mathematica* **2** (1975) 529–533.
39. D. Howard, "Einstein and the Development of Twentieth-Century Philosophy of Science", Chapter 11 in: M. Janssen and C. Lehner (eds), The Cambridge Companion to Einstein, Cambridge University Press (2014) pp. 354–376.
40. A. Einstein, "Physics and reality", translated by J. Piccard, The Journal of the Franklin Institute 221 (1936) 3349–382.
41. J.O. Urmson and G.J. Warnock (eds), *J. L. Austin; Philosophical Papers*, Oxford University Press (1979) p. 252.
42. FXQi discussion forum for the current essay, URL: http://fqxi.org/community/forum/topic/2492 (retrieved Sept. 2015).

Mathematics Is Physics

M. S. Leifer

Abstract In this essay, I argue that mathematics is a natural science—just like physics, chemistry, or biology—and that this can explain the alleged "unreasonable" effectiveness of mathematics in the physical sciences. The main challenge for this view is to explain how mathematical theories can become increasingly abstract and develop their own internal structure, whilst still maintaining an appropriate empirical tether that can explain their later use in physics. In order to address this, I offer a theory of mathematical theory-building based on the idea that human knowledge has the structure of a scale-free network and that abstract mathematical theories arise from a repeated process of replacing strong analogies with new hubs in this network. This allows mathematics to be seen as the study of regularities, within regularities, within..., within regularities of the natural world. Since mathematical theories are derived from the natural world, albeit at a much higher level of abstraction than most other scientific theories, it should come as no surprise that they so often show up in physics. This version of the essay contains an addendum responding to Slyvia Wenmackers' essay [1] and comments that were made on the FQXi website [2].

The Unreasonable Effectiveness of Mathematics in the Physical Sciences

> The miracle of the appropriateness of the language of mathematics for the formulation of the laws of physics is a wonderful gift which we neither understand nor deserve—Eugene Wigner [3].

Mathematics is the language of physics, and not just any mathematics at that. Our fundamental laws of physics are formulated in terms of some of the most advanced and abstract branches of mathematics. Seemingly abstruse ideas like differential geometry, fibre bundles, and group representations have been commonplace in physics for decades, and theoretical physics is only getting more abstract, with fields like

M.S. Leifer (✉)
Perimeter Institute for Theoretical Physics, 31 Caroline Street North,
Waterloo, ON N2L 2Y5, Canada
e-mail: matt@mattleifer.info

A. Aguirre et al. (eds.), *Trick or Truth?*, The Frontiers Collection,
DOI 10.1007/978-3-319-27495-9_3

category theory playing an increasingly important role in our theory building. Do we simply have to accept this mathematization as a brute fact about our universe, or can it be explained?

My thesis is that mathematics is a natural science—just like physics, chemistry, or biology—albeit one that is separated from empirical data by several layers of abstraction. It is nevertheless fundamentally a theory about our physical universe and, as such, it should come as no surprise that our fundamental theories of the universe are formulated in terms of mathematics.

The philosophical worldview underlying my arguments is that of naturalism [4, 5]. Naturalism is the position that everything arises from natural properties and causes, i.e. supernatural or spiritual explanations are excluded. In particular, natural science is our best guide to what exists, so natural science should guide our theorizing about the nature of mathematical objects.

Mathematics is not usually thought of as an empirical science, so naturalism may seem irrelevant to its philosophy. However, we have one rather conspicuous empirical data point about it, namely the alleged "unreasonable" effectiveness of mathematics in the physical sciences. Since our fundamental laws of physics are formulated in terms of some of the most advanced branches of mathematics, a philosophy of mathematics that explains this should be preferred to one in which it is an "unreasonable" accident or miracle. Such a theory would also be falsifiable in the sense that, were it to be the case our future fundamental theories of physics resist mathematization, being only explainable in words or only formalizable in terms of very elementary mathematics, then our philosophy of mathematics would have been proved wrong.

The philosophy of mathematics I develop here is closely related to those of Quine and Putnam [6–9], who instigated the naturalistic approach to mathematics and even suggested that logic could be empirical. However, unlike them, I am inclined towards a more pragmatic theory of truth so, for me, abstract mathematical objects can be called real insofar as they are useful for our scientific reasoning. This evades the problem of trying to find direct referrents of mathematical objects in the physical world. Instead, I simply have to explain why they are useful. In order to do this, I shall have to investigate where mathematics comes from. In this vein, I shall argue that human knowledge has the structure of a *scale-free network* and offer a theory of mathematical theory-building that emphasizes reasoning by analogy within this network. This explains how mathematics can become increasingly abstract, whilst maintaining its tether to empirical reality. It also explains why abstract areas of mathematics that are developed in seeming isolation from physics often show up later in our physical theories.

My title, "Mathematics is Physics", is deliberately chosen in contrast to our dear leader Max Tegmark's Mathematical Universe Hypothesis [10] (see Fig. 1 for a comparison of the two views). This asserts that our universe is nothing but a mathematical structure and that all possible mathematical structures exist in the same sense as our universe. The first part of the hypothesis may be condensed to "Physics is Mathematics", so in this sense I am arguing for precisely the opposite. I will compare and contrast the two ideas at the end.

Fig. 1 **a** The staff at my local bookstore agree with Tegmark's Mathematical Universe Hypothesis, as they have displayed his book about the nature of our physical universe in the mathematics section. **b** In response, I placed a mathematics book in the physics section, in order to illustrate my counterhypothesis

What Is Mathematics?

The philosophy of mathematics is now a vast and sprawling subject,[1] so I cannot possibly cover all views of the nature of mathematics here. However, it is useful to introduce two of the more traditional schools of thought—mathematical platonism and formalism—to serve as foils to the naturalistic theory I then develop.

Mathematical Platonism

Mathematical platonism is the idea that mathematical objects are abstract entities that exist objectively and independently of our minds and the physical universe. For example, the geometric concept of a straight line does not refer to any approximation of a straight line that we might draw with pencil on paper. If we view any real world approximation through a microscope we will see that it has rough edges and a finite thickness. On the other hand, a geometric straight line is perfectly straight and has no thickness. It is a fundamentally one-dimensional object, which we can only ever approximate in our three-dimensional universe. Mathematical objects like straight

[1]See [11] for an overview of the subject.

lines do not exist in our universe, so the platonist asserts that they exist in an abstract realm, somewhat akin to the Platonic world of forms. They further assert that we have direct access to this realm via our mathematical intuition.

Mathematical platonism is in direct conflict with naturalism. A naturalistic theory has no place for a dualistic mind that is independent of the structure of our brains. Therefore, if we have intuitive access to an abstract realm, our physical brains must interact with it in some way. Our best scientific theories contain no such interaction.[2] The only external reality that our brains interact with is *physical* reality, via our sensory experience. Therefore, unless the platonist can give us an account of where the abstract mathematical realm actually is in physical reality, and how our brains interact with it, platonism falls afoul of naturalism.

Furthermore, our brains are the product of evolution by natural selection, so, on the naturalist view, whatever mathematical intuitions we have are either a reflection of what was useful for survival, or products of the general intelligence that evolution has endowed us with. Our evolutionary instincts are often poor guides to reality, so it is difficult to see how mathematics could be objective if mathematical intuition is of this type. If it is instead a product of general intelligence then mathematics could either be akin to a creative work of fiction, or it must be the result of reasoning about the physical world that we find ourselves in. Only the latter can explain why our theories of physics are mathematical, so this account should be preferred.

Formalism

Formalism is the view that mathematics is just a game about the formal manipulation of symbols. We specify some symbols, such as marks on a sheet of paper, and rules for deriving one string of symbols from another. Anything that can be derived from those rules is a theorem of the resulting mathematical system.

As a toy example, we could posit that the symbols are 0 and 1. The rules are that you may replace the empty string with either 0 or 1, you may replace any string s ending in 0 with $s1$, and any string s ending in 1 with $s0$. $10 \rightarrow 1010101$ is an example of a (very boring) theorem in this (very boring) formal system.

Formalism has the advantage that it untethers mathematics from an abstract objective realm that is independent of our minds. It is therefore naturalistic in the sense that it does not posit a supernatural world. Mathematics instead becomes an intellectual game, in which we may posit any rules we like. However, formalism has two difficult obstacles to deal with. Firstly, mathematicians do not study arbitrary formal systems. There is no "adding zeroes and ones to the end of binary strings" research group in any mathematics department. Formalists must specify which formal systems count as mathematics. Why is group theory a branch of mathematics, but adding zeroes

[2]This argument is known as Benacerraf's epistemological objection to mathematical platonism [12].

and ones to the end of binary strings is not? Secondly, even if this is achieved, the formalist has no explanation of why abstract branches of mathematics show up in physics.

Mathematics as a Natural Science

In response to the objections to platonism and formalism, I wish to defend the idea that mathematics is a natural science, i.e. its subject matter is ultimately our physical universe. To do this, I will borrow one idea each from platonism and formalism. Along with the platonists, I want to view mathematics as being about an objective world that exists independently of us. The difference is that, in my case, this is just the actual physical world that we live in, rather than some hypothetical abstract world of forms.

However, mathematical objects are more abstract than those that appear in the physical world, and they include entities that seem to have no physical referrent, such as hierarchies of infinities of ever increasing size. To deal with this issue, I maintain that mathematical objects do not refer directly to things that exist in the physical universe. As the formalists suggest, mathematical theories are just abstract formal systems, but not all formal systems are mathematics. Instead, mathematical theories are those formal systems that maintain a tether to empirical reality through a process of abstraction and generalization from more empirically grounded theories, aimed at achieving a pragmatically useful representation of regularities that exist in nature.

It is relatively easy to defend the idea that elementary concepts like the finite natural numbers are theories of things that actually exist in the world, viz. the rules of arithmetic are derived from what actually happens if you combine collections of discrete objects such as sheep, rocks, or apples. However, the advanced theories of mathematics deal with entities that are not related to the physical world in any obvious way at all. For example, nobody has ever seen an infinite collection of sheep of any type, let alone one that has the cardinality of some specific transfinite number. Moreover, mathematical theories seem to have their own autonomy, independent of the natural sciences. Pure mathematicians have their own programs of research, with well-motivated questions that are internal to their mathematical theories, making no obvious reference to empirical reality. In response to this, some naturalists are content to only deal with the mathematics that actually occurs in science, and it has been suggested that the theory of functions of real variables is sufficient for this. However, much more abstract mathematics than this appears in modern physics and, in any case, if we are to explain why the fruits of pure mathematics research so often appear in physics at a later date, then we are going to need a theory that encompasses purely abstract research. So, for me, the biggest problem is to explain how abstract pure mathematical theories can be empirical theories in disguise, and to do so in such a way that the later application of these theories to physics becomes natural.

This question cannot be answered without looking at where mathematical theories come from. If I can argue that mathematical theories maintain an appropriate tether to empirical reality then it should be no surprise that the regularities encoded within them, which already refer to nature, later show up in natural science. Developing a theory of human knowledge and mathematical theory-building that does this is the main challenge for my approach.

The Structure of Human Knowledge

Although we are concerned with how mathematics relates to the physical world, it is important to realize that all our knowledge is *human* knowledge, i.e. it is discovered, organized, learned, and evaluated in an ongoing process by a social network of finite beings. This means that the structure of our knowledge will, in addition to reflecting the physical world, also reflect the nature of the process that generates it. By this I do not mean to imply that human knowledge is merely a social construct—it is still knowledge about an objective physical world. However, if we are to explain why abstract pure mathematics later shows up in physics, we are going to have to examine the motivations and methodology of those who develop that mathematics. We have to uncover a hidden empirical tether in their methods and explain how it can be that the patterns and regularities they are studying are in fact patterns and regularities of nature in disguise.

It is common to view the structure of human knowledge as hierarchical. The various attempts to reduce all of mathematics to logic or arithmetic reflect a desire view mathematical knowledge as hanging hierarchically from a common foundation. However, the fact that mathematics now has multiple competing foundations, in terms of logic, set theory or category theory, indicates that something is wrong with this view.

Instead of a hierarchy, we are going to attempt to characterize the structure of human knowledge in terms of a network consisting of nodes with links between them (see Fig. 2). Roughly speaking, the nodes are supposed to represent different fields of study. This could be done at various levels of detail. For example, we could draw a network wherein nodes represent things like "physics" and "mathematics", or we could add more specific nodes representing things like "quantum computing" and "algebraic topology". We could even go down to the level having nodes representing individual concepts, ideas, and equations. I do not want to be too precise about where to set the threshold for a least digestible unit of knowledge, but to avoid triviality it should be set closer to the level of individual concepts than vast fields of study.

Next, a link should be drawn between two nodes if there is a strong connection between the things they represent. Again, I do not want to be too precise about what this connection should be, but examples would include an idea being part of a wider theory, that one thing can be derived from the other, or that there exists a strong direct analogy between the two nodes. Essentially, if it has occurred to a human being that the two things are strongly related, e.g. if it has been thought interesting enough to

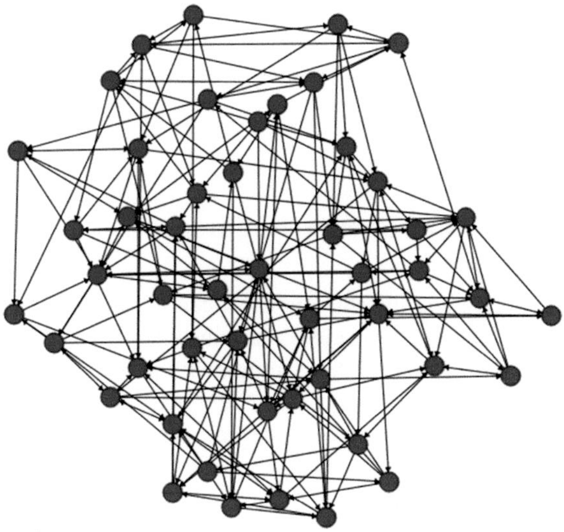

Fig. 2 An example of a network with nodes (*red circles*) and links between them (*arrows*). This network was generated by the SocNetV software package [13]

do something like publish an academic paper on the connection, and the connection has not yet been explained in terms of some intermediary theory, then there should be a link between the corresponding nodes in the network.

If we imagine drawing this network for all of human knowledge then it is plausible that it would have the structure of a *scale-free network* [14]. Without going into technical details, scale-free networks have a small number of *hubs*, which are nodes that are linked to a much larger number of nodes than the average. This is a bit like the 1 % of billionaires who are much richer than the rest of the human population. If the knowledge network is scale-free then this would explain why it seems so plausible that knowledge is hierarchical. In a university degree one typically learns a great deal about one of the hubs, e.g. the hub representing fundamental physics, and a little about some of the more specialized subjects that hang from it. As we get ever more specialized, we typically move away from our starting hub towards more obscure nodes, which are nonetheless still much closer to the starting hub than to any other hub. The local part of the network that we know about looks much like a hierarchy, and so it is not surprising that physicists end up thinking that everything boils down to physics whereas sociologists end up thinking that everything is a social construct. In reality, neither of these views is right because the global structure of the network is not a hierarchy.

As a naturalist, I should provide empirical evidence that human knowledge is indeed structured as a scale-free network. The best evidence that I can offer is that the structure of pages and links on the World Wide Web and the network of citations to academic papers are both scale free [14]. These are, at best, approximations of the true knowledge network. The web includes facts about the Kardashian family that I do not want to categorize as knowledge, and not all links on a website indicate a strong connection, e.g. advertising links. Similarly, there are many reasons why people cite papers other than a direct dependence. However, I think that these examples provide evidence that the information structures generated by a social network of finite beings are typically scale-free networks, and the knowledge network is an example of such a structure.

A Theory of Theory-Building

We are now at the stage where I can explain where I think mathematical theories come from. The main idea is that when a sufficiently large number of strong analogies are discovered between existing nodes in the knowledge network, it makes sense to develop a formal theory of their common structure, and replace the direct connections with a new hub, which encodes the same knowledge more efficiently.

As a first example, consider the following just-so story about where natural numbers and arithmetic might have come from. Initially, people noticed that discrete quantities of sheep, rocks, apples, etc. all have a lot of properties in common. Absent a theory of this common structure, the network of knowledge has a vast number of direct connections between the corresponding nodes (see Fig. 3). It therefore makes sense to introduce a more abstract theory that captures the common features of all these things, and this is where the theory of number comes in. A vast array of individual connections is replaced by a new hub, which has the effect of organizing knowledge more efficiently. Now, instead of having to learn about quantities of sheep, rocks, apples, etc. individually and then painstakingly investigate each analogy, one need only learn about the theory of number and then apply it to each individual case as needed. In this way, the theory of number remains essentially empirical. It is about regularities that exist in nature, but is removed from the empirical data by one layer of abstraction compared to our direct observations.

Once it is established, the theory of number allows for the introduction of new concepts that are not present in finite collections of sheep, such as infinite sequences and limits. As the theory is more abstract than the empirical phenomena it is derived from, it develops its own internal life and is partially freed from its empirical ties. Such internal questions are sometimes settled by pragmatic considerations, e.g. which definitions make for the most usable extension of the theory beyond strictly finite quantities, but mostly by the way in which the theory ends up interacting with the larger structure of mathematical knowledge. For example, the question of how to define limits of infinite sequences was not really settled until those definitions were needed to understand calculus and analysis, which are themselves abstractions of the physically rooted geometries and flows that exist in our universe.

Once several abstract theories have been developed, the process can continue at a higher level. For example, category theory was born out of the strong analogies that exist between the structure preserving maps in group theory, algebraic topology, and homological algebra [15]. At first sight, it seems like this is a development that is completely internal to pure mathematics, but really what is going on is that mathematicians are noticing regularities, within regularities, within regularities…, within

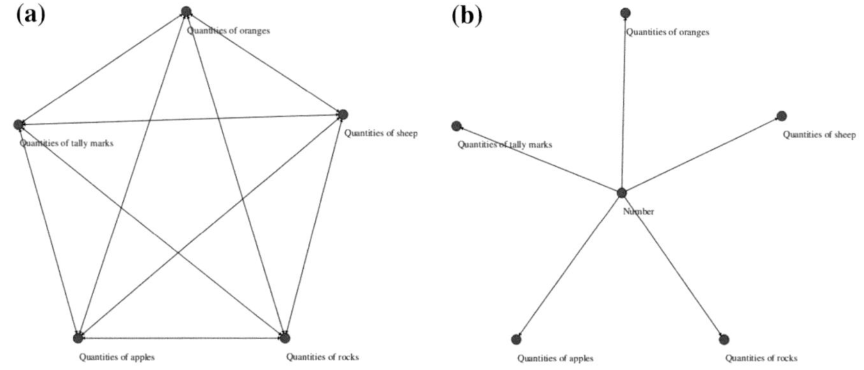

Fig. 3 How abstract mathematical theories arise from analogies between theories at a lower level of abstraction, which are closer to the empirical data. **a** Absent a theory of number, there are a large number of direct analogies between quantities of discrete objects like sheep, rocks, apples, etc. Only analogies between five types of discrete objects are depicted here. There would be vastly more in the real knowledge network. **b** A new hub is introduced to capture the common structure indicated by the analogies. The theory of number captures the regularity at one higher level of abstraction

regularities of the physical world. In this way, mathematics can become increasingly abstract, and develop its own independent structure, whilst maintaining a tether to the empirical world.

Now, to be sure, in the course of this process mathematicians have to make several pragmatic choices that seem to be independent of physics. For example, is it better to reject the axiom of choice, which says that when you have several sets of objects (including a possibly infinite number of sets) then there is a way of picking one object from each of them, or is it better to accept the counter-intuitive implication of this that if you peel a mathematically idealized orange then the peel can be used to completely cover two mathematically idealized oranges of the same size?[3] Ultimately, the axiom of choice is accepted by most mathematicians because it leads to the most useful theories. The theories that employ it lead to a more efficient knowledge network than those that reject it, and this trumps any apparent physical implausibility. In my view, intuition for efficient knowledge structure, rather than intuition about an abstract mathematical world, is what mathematical intuition is about.

It is now straightforward to explain why abstract mathematical theories show up so often in physics. Abstract mathematical theories are about regularities, within regularities, within…, within regularities of our physical world. Physical theories are about exactly the same thing. The only difference is that whilst mathematics started from empirical facts that only required informal observations, physics includes the much more accurate empirical investigations that only became possible due to scientific and technological advances, e.g. the development of telescopes and particle colliders. Nonetheless, it should be no surprise that regularities that are about the selfsame physical world turn out to be related. It should also be no surprise that

[3]This is the Banach–Tarski paradox [16].

pure mathematicians often develop the mathematics relevant for physics a long time before it is needed by physicists, as they have had much more time to investigate regularities at a higher level of abstraction.

Implications

Before concluding, I want to describe two implications of my theory of knowledge and mathematics for the future of physics.

Firstly, in network language, the concept of a "theory of everything" corresponds to a network with one enormous hub, from which all other human knowledge hangs via links that mean "can be derived from". This represents a hierarchical view of knowledge, which seems unlikely to be true if the structure of human knowledge is generated by a social process. It is not impossible for a scale-free network to have a hierarchical structure like a branching tree, but it seems unlikely that the process of knowledge growth would lead uniquely to such a structure. It seems more likely that we will always have several competing large hubs and that some aspects of human experience, such as consciousness and why we experience a unique present moment of time, will be forever outside the scope of physics.

Nonetheless, my theory suggests that the project of finding higher level connections that encompass more of human knowledge is still a fruitful one. It prevents our network from having an unwieldy number of direct links, allows us to share more common vocabulary between fields, and allows an individual to understand more of the world with fewer theories. Thus, the search for a theory of everything is not fruitless; I just do not expect it to ever terminate.

Secondly, my theory predicts that the mathematical representation of fundamental physical theories will continue to become increasingly abstract. The more phenomena we try to encompass in our fundamental theories, the further the resulting hubs will be from the nodes representing our direct sensory experience. Thus, we should not expect future theories of physics to become less mathematical, as they are generated by the same process of generalization and abstraction as mathematics itself.

Conclusions

I have argued that viewing mathematics as a natural science is the only reasonable way of understanding why mathematics plays such a central role in physics. I have also offered a theory of mathematical theory-building that can explain how mathematical theories maintain a tether to the physical world, despite becoming ever more abstract.

To conclude, I want to contrast my theory with Tegmark's Mathematical Universe Hypothesis. In my theory, abstract mathematics connects to the physical world via our direct empirical observations. The latter are at the very edges of the knowledge network, as far away from our most abstract mathematical theories as they possibly could be. They are the raw material from which our mathematical theories are constructed, but the theories themselves are just convenient representations of the regularities, within regularities, within..., within regularities of the physical world. Thus, my view is opposite to Tegmark's. Mathematics is constructed out of the physical world rather than the other way round.

Like my proposal, the Mathematical Universe Hypothesis is naturalistic. It asserts that there is no abstract mathematical realm independent of our physical universe, because the physical world is *identified* with the realm of all possible mathematical structures. However, the universe we find ourselves in is just one structure in a multiverse of equally real possibilities, and Tegmark's theory does not explain how we could come to know about these other mathematical structures. Indeed, the universes within a multiverse should not have a strong interaction with each other, otherwise they would not be identifiable as independent universes, so it is difficult to see how there could be any causal connection to explain our abstract mathematical theorizing.

The main evidence for the Mathematical Universe Hypothesis is that physics is becoming ever more abstract and mathematical, so it looks like the world described by physics can be identified with a mathematical structure. However, my theory provides an explanation for the increasing abstraction of physics, and can also account for mathematical theory-building in a natural way so, at present, I think it should be preferred.

Addendum

I have been asked to update this essay in light of the discussion that occurred on the FQXi website during the competition. I cannot possibly cover all of the comments, so I shall focus on a few that I consider interesting and accessible. For more details, it is worth reading the comment thread in its entirety [2].

Since Sylvia Wenmackers' winning essay also argues for a naturalistic approach to mathematics [1], I also thought it worthwhile to discuss how my position differs from hers. We start with that before moving on to the website comments.

Response to Sylvia Wenmackers' Essay

The main advantage of a naturalistic view of mathematics is that it offers a simple explanation of the alleged unreasonable effectiveness of mathematics in physics. If mathematics is fundamentally about the physical world then it is no surprise that it occurs in our description of the physical world.

However, the naturalistic view is prima facie absurd because mathematical truth seems to have nothing to do with physical world. The main task of a naturalistic account of mathematics is therefore a deflationary one: explain why mathematical theories are empirical theories in disguise and why we have been so easily misled into thinking this is not the case. Wenmackers gives a battery of arguments for this position based on four "elements". Here, I will focus on the first and third elements, which are based on evolution by natural selection and selection bias respectively, because this is where I disagree with her. I more or less agree with her discussion of the other two elements.

The Evolutionary Argument

The evolutionary argument asserts that our cognitive abilities, and in particular our ability to do mathematics, are the result of evolution by natural selection. It is therefore no surprise that they reflect physical reality, as physical reality provided the environmental pressures that selected for those abilities.

This argument works well for basic mathematics, such as arithmetic and elementary pattern recognition. The ability to distinguish a tree that has five apples growing on it from one that has two has an obvious evolutionary advantage.

However, our main task is to explain why our most advanced and abstract theories of mathematics crop up so often in modern physics, not just the basic theories like arithmetic, and there is no conceivable evolutionary pressure towards understanding the cosmos on a large scale. Knowing the ultimate fate of the universe may well be crucial for our (very) long term survival, but cosmology operates on a much longer timescale than evolution by natural selection, so, for example, there is no immediate environmental pressure towards discovering general relativity, nor the differential geometry needed to formulate it.

On the other hand, it happens that evolution has endowed us with general curiosity and intelligence. This does have a survival advantage as, for example, a species that is able to detect patterns in predator attacks and pass that knowledge on to the next generation without waiting for genetic changes to make the knowledge innate can adapt to its environment more quickly. Such curiosity and intelligence are not the inevitable outcome of evolution, as the previous dominance of dinosaurs on this planet demonstrates, but just one possible adaptation that happened to occur in our case. Like many adaptations, it has side effects that are not immediately related to our survival, one of which is that some of us like to think about the large scale cosmos.

Since our general curiosity and intelligence are only a side effect of adaptation, so are modern physics and mathematics. Therefore, I find it puzzling to argue for the efficacy of our reasoning in these areas based on evolution. Evolution often endows beliefs and behaviours that are good heuristics for the cases commonly encountered, but a poor reflection of reality (consider optical illusions for example). With general intelligence one can, with considerable effort, reason oneself out of such beliefs and behaviours. This is why I stated that mathematical intuition must be a product of general intelligence rather than a direct evolutionary adaptation in my essay.

I think that understanding how a network of intelligent beings go about organizing their knowledge is at the root of the efficacy of mathematics in physics. It should not matter whether those beings are the products of evolution by natural selection or some hypothetical artificial intelligences that we may develop in the future. For this reason,

I take the existence of intelligent beings as a starting point, rather than worrying about how they got that intelligence.

Selection Bias

Wenmackers' selection bias argument is an attempt to deflate the idea that mathematics is unreasonably effective in physics. The idea is that, due to selection bias, we tend to remember and focus on those cases in which mathematics was successfully applied in physics, whereas the vast majority of mathematics is actually completely useless for science. Here is the argument in her own words.

> Among the books in mathematical libraries, many are filled with theories for which not a single real world application has been found. We could measure the efficiency of mathematics for the natural science as follows: divide the number of pages that contain scientifically applicable results by the total number of pages produced in pure mathematics. My conjecture is that, on this definition, the efficiency is very low. In the previous section we saw that research, even in pure mathematics, is biased towards the themes of the natural sciences. If we take this into account, the effectiveness of mathematics in the natural sciences does come out as unreasonable - unreasonably low, that is.

My first response to this is to question whether the efficiency is actually all that low. After all, the vast majority of pages written by theoretical physicists might also be irrelevant to reality, and these are people who are deliberately trying to model reality. We need only consider the corpus of mechanical models of the ether from the 1800s to render this plausible, let alone the vast array of current speculative models of cosmology, particle physics, and quantum gravity. It is not obvious to me whether the proportion of applicable published mathematics is so much lower than the proportion of correct published physics, and, if it is not, then a raw page count does not say much about the applicability of mathematics in particular.

Even if the efficiency of mathematics is much lower than that of physics, it not obvious how low an inefficiency would be unreasonably low. If mathematics were produced by monkeys randomly hitting the keys of typewriters then the probability of coming up with applicable mathematics would be ridiculously small, akin to a Boltzmann brain popping into existence via a fluctuation from thermal equilibrium. In light of such a ridiculously tiny probability, an efficiency of say 0.01 %, which looks small from an everyday point of view, would indicate an extremely high degree of unreasonable effectiveness. Of course, mathematicians are not typewriting monkeys, but unless one is already convinced that the development of mathematics is correlated with the development physics by one of Wenmackers' other arguments, then even a relatively tiny efficiency could seem extremely improbable.

My second response to the selection bias argument is that mathematics is not identical to the corpus of mathematical literature laid out in a row. Some mathematical theories are considered more important than others, e.g. the core topics taught in an undergraduate mathematics degree. Therefore, mathematical theories ought to be weighted with their perceived importance when calculating the efficiency of mathematics. If you buy my network model of knowledge then the number of inbound links to a node could be used to weight its importance, as in the Google Page rank algorithm. I would conjecture that the efficiency of mathematics is much higher when weighted by perceived importance. I admit that this argument could be accused of circularity because one of the reasons why an area of mathematics might be regarded as important is its degree of applicability. However, this just reinforces the point that mathematics not an isolated subject, but must be considered in the context of the whole network of human knowledge.

Responses to Comments on the FQXi Website

Other Processes in the Knowledge Network

Several commenters expressed doubts that my theory of theory building captures everything that is going on in mathematics. For example, Wenmackers commented:

> Is this is correct summary of your main thesis (in Section 4)? : "First, humans studied many aspects the world, gathering knowledge. At some point, it made sense to start studying the structure of that knowledge. (And further iterations.) This is called mathematics."
>
> Although I find this idea appealing (and I share your general preference for a naturalistic approach), it is not obvious to me that this captures all (or even the majority) of mathematical theories. In mathematics, we can take anything as a source of inspiration (including the world, our the structure of our knowledge there of), but we are not restricted to studying it in that form: for instance, we may deliberately negate one of the properties in the structure that was the original inspiration, simply because we have a hunch that doing so may lead to interesting mathematics. Or do you see this differently?

There are other processes going on in the knowledge network beyond the theory-building process that I described in my essay. I did not intend to suggest otherwise. The reason why I focused on the process of replacing direct links by more abstract theories is because I think it can explain how mathematics becomes increasingly abstract, whilst maintaining its applicability. But this is clearly not the only thing that mathematicians do.

One additional process that is going on is a certain amount of free play and exploration, as also noted by Ken Wharton in his essay [17]. Mathematical axioms may be modified or negated to see whether they lead to interesting theories. However, as I argued earlier, mathematical theories should be weighted with their perceived importance when considering their place in the corpus of human knowledge. Modified theories that are highly connected to existing theories, or to applications outside of mathematics, will ultimately be regarded as more important. It is possible that a group of pure mathematicians will end up working on a relative backwater for an academic generation or two, but this is likely to be given up if no interesting connections are forthcoming.

For my theory, it is important that these additional processes should not have a significant impact on the broad structure of the knowledge network. There should not be a process where large swaths of pure mathematicians are led to work on completely isolated areas of the network, developing a large number of internal links that raise the perceived importance of their subject, with almost no links back to the established corpus of knowledge. Personally, I think that any such process is likely to be dominated by processes that do link back strongly to existing knowledge, but this is an empirical question about how the mathematical knowledge network grows. To address it, I would need to develop concrete models, and compare them to the actual growth of mathematics.

What Physical Fact Makes a Mathematical Fact True?

Perhaps the highlight of the comment thread was a discussion with Tim Maudlin. It started with the following question:

> I'm not sure I understand the sense in which mathematics is supposed to be "about the physical world" as you understand it. In one sense, the truth value of any claim about the physical world depends on how the physical world is, that is, it is physically contingent. Had the physical world been different, the truth value of the claim would be different. Now take a claim about the integers, such as Goldbach's conjecture. Do you mean to say that the truth or falsity of Goldbach's conjecture depends on the physical world: if the physical world is one way then it is true and if it is another way it is false? What feature of the physical world could the truth or falsity of the conjecture possibly depend on?

I stated in the essay that I think mathematical theories are formal systems, but not all formal systems are mathematics. The role of physics is to help delineate which formal systems count as mathematics. Therefore, if Goldbach's conjecture is a theorem of Peano arithmetic then I would say it is true in Peano arithmetic. If we want to ask whether it is true of the world then we have to ask if Peano arithmetic is the most useful version of arithmetic by looking how it fits into the knowledge network as a whole. It may be that more than one theory of arithmetic is equally useful, in which case we may say that it has indefinite truth value, or we may want to say that it is true in one context and not another if the two theories of arithmetic have fairly disjoint domains of applicability.

If the Goldbach conjecture is not provable in Peano arithmetic, but is provable in some meta-theory, then we can ask the same questions at the level of the meta-theory, i.e. is it more useful than a different meta-theory. This sort of consideration has happened in mathematics in a few cases, e.g. most mathematicians choose to work under the assumption that the axiom of choice is true, mainly, I would argue, because it leads to more useful theories.

At this point in the discussion, Maudlin accused me of being a mathematical platonist. His point is that if I accept theoremhood as my criterion of truth then I am admitting some mathematical intuitions as self-evident truths, so if my goal is to remove intuition as the arbiter of mathematical truth then I have not yet succeeded. Specifically, in order to even state what it means for something to be a theorem in a formal system, I need to accept at least some of the structure informal logic, e.g. things like: if A is true and B is true then A AND B is also true. Why do we accept these ideas of informal logic? Primarily because they seem to be self-evidently true, but this is an appeal to unfettered mathematical intuition.

Maudlin points out that, because of this, it is difficult to avoid some form of mathematical platonism, if only about the basic ideas of logic. I am not yet prepared to accept this and would argue, along with Quine and Putnam [6, 8], that logic may be empirical (in my case, I would say that even very basic informal logic may be empirical).

I do not deny that I accept informal logic because it seems self-evident to me. However, as I argued in the essay, my intuitions come from my brain, and my brain is a physical system. Therefore, if I have strong intuitions, they must have been put there by the natural processes that led to the development of my brain. The likely culprit in this case is evolution by natural selection. Organisms that intuitively accept the laws of informal logic survive better than those that do not because those laws are true of our physical universe, so that creates a selection pressure to have them built in as intuitions.

Does this mean there are conceivable universes in which the laws of basic logic are different,[4] e.g. in which Lewis Carroll's modus ponens denying tortoise [18] is correct. I have to admit that I have difficulty imagining it, but it does not seem totally inconceivable. In such a universe, what counts as a mathematical truth would be different. In other words, mathematical truth might be empirical because the laws of logic are.

In addition to this, there is a much more prosaic way in which mathematical truth is dependent on the physical laws. Imagine a universe in which planets do not make circular motions as time progresses, but instead travel in straight lines through a continually changing landscape. Inhabitants of such a planet would probably not measure time using a cyclical system of minutes, hours, days, etc. as we do, but instead just use a system of monotonically increasing numbers. A bit more fancifully, suppose that in this hypothetical universe, every time a collection of twelve discrete objects like sheep, rocks, apples, etc. are brought together in one place they magically disappear into nothingness. Inhabitants of this world would use clock arithmetic, technically known as mod 12 arithmetic, to describe collections of discrete objects. Their view of how to measure time versus how to measure collections of discrete objects would be precisely the reverse of ours. At least one of the senses in which $12 + 1 = 13$ in our universe is not true in theirs, and I would say that this is a sense in which mathematical truth depends on the laws of physics.

It is fair to say that Maudlin was not impressed by this example, but I take it deadly seriously. What counts as mathematics and what counts as mathematical truth are, in my view, pragmatically dependent on how our mathematical theories fit into the structure of human knowledge. If the empirical facts change then so does the structure of this network. The meaning of numbers, in particular, is dependent on how collections of discrete objects behave in our universe, and if you change that

[4] The usual terminology for hypothetical universes with different laws is "logically possible". That seems inappropriate here, but the terminology does show how ingrained into our minds the basic laws of logic are.

then you change what makes a given theory of number useful, and hence true in the pragmatic sense. It is this that makes the theory of number a hub in the network of human knowledge and this is what philosophers ought to be studying if they want to understand the meaning of mathematics. The usual considerations in the foundations of mathematics, such as deriving arithmetic from set theory, though still well-connected to other areas of mathematics, are comparative backwaters. If we really want to understand what mathematics is about, we ought to get our heads out of the formal logic textbooks and look out at the physical world.

Why are There Regularities at All?

Sophia Magnusdottir points out that my approach does not address why there are regularities in nature to begin with.

> In a nutshell what you seem to be saying is that one can try to understand knowledge discovery with a mathematical model as well. I agree that one can do this, though we can debate whether the one you propose is correct. But that doesn't explain why many of the observations that we have lend themselves to mathematical description. Why do we find ourselves in a universe that does have so many regularities? (And regularities within regularities?) That really is the puzzling aspect of the "efficiency of mathematics in the natural science". I don't see that you address it at all.

There are two relevant kinds of regularities here: the regularities described by our most abstract mathematical theories on the one hand, and the regularities of nature on the other. On the face of it, these two types of regularity have little to do with one another. The fact that the regularities described by our most abstract mathematical theories so often show up in physics is what I take to be the "unreasonable" effectiveness of mathematics in the physical sciences.

What I have tried to do is to argue that these two types of regularity are more closely connected than we normally suppose. They both ultimately describe regularities, within regularities, ..., within the natural world. I have not even tried to address the question of why there are regularities in nature in the first place. Instead, I have taken their existence as my starting point. If we live in a universe with regularities, the process of knowledge growth is such that what we call mathematics will naturally show up in physics. This answers what I take to be the problem of "unreasonable" effectiveness.

Of course, one can try to go further by asking why there are any regularities in the first place. I do not think that anyone has provided a compelling answer to this, and I suspect that it is one of those questions that just leads to an infinite regress of further "why" questions.

For example, the Mathematical Universe Hypothesis may seem, superficially, to explain the existence of regularities. If our universe literally is a mathematical structure, and mathematical structures describe regularities, then there will necessarily be regularities in nature. However, one can then ask why our universe is a mathematical structure, which is just the same question in a different guise.

If we are to take the results of science seriously, the idea that our universe is sufficiently regular to make science reliable has to be assumed. There is no proof of this, despite several centuries of debate on the problem of induction. Although this is an interesting issue, I doubt that the problem can ever be resolved in an uncontroversial way and it seems, to me at least, to be a different and far more difficult problem than the "unreasonable" effectiveness of mathematics in physics. If my ideas are correct then at least there is now only one type of regularity that needs to be explained.

Elegance or Efficiency?

Alexy Burov made the following point.

> Wigner's wonder about the relation of physics and mathematics is not just abut the fact that there are some mathematical forms describing laws of nature. He is fascinated by something more: that these forms are both elegant, while covering a wide range of parameters, and extremely precise. I do not see anything in your paper which relates to that amazing and highly important fact about the relation of physics and mathematics.

I take the key issue here to be that I have not explained why the mathematics used in physics is "elegant". After all, if we had a bunch of different laws covering different parameter ranges then we could always put them together into a single structure by inserting a lot of "if" clauses into our laws of physics. We can also make them arbitrarily precise by adding lots of special cases in this way. Presumably though, the result of this would be judged "inelegant".

To be honest, I have a great deal of trouble understanding what mathematicians and physicists mean by "elegance" (hence the scare quotes). For this reason, I have emphasized that the mathematics in modern physics is "abstract" and "advanced" rather than "elegant".

A more precise definition of elegance is needed to make any progress on this issue. One concrete suggestions is that perhaps elegance refers to the fact that the fundamental laws of physics are few in number so they can be written on a t-shirt. It is tempting to draw the analogy with algorithmic information here, i.e. the length of the shortest computer program that will generate a given output [19]. Perhaps the laws of physics are viewed as elegant because they have low algorithmic information. We get out of them far more than we put in.

So, perhaps what we call "elegance" really means the smallest possible set of laws that encapsulates the largest number of phenomena. If so, then what we need to explain is why the process of scientific discovery would tend to produce laws with low algorithmic information. The idea that scientists are trying to optimize algorithmic information directly is a logician's parody of a complex social process. Instead, we need to determine whether the processes going on in the knowledge network would tend to reduce the algorithmic information content of the largest hubs in the network. In this I am encouraged by the fact that many scale-free networks exhibit the "small world" phenomenon in which the number of links in a path connecting two randomly chosen nodes is small.[5] If this is true of the knowledge network then it means that the hubs must be powerful enough to derive the empirical phenomena in a relatively small number of steps. The average path length between two nodes might be taken as a measure of the efficiency with which our knowledge is encoded in the network, or, if you prefer, its "elegance".

Now, of course, this may be completely unrelated to what everyone else means by the word "elegance", as applied to mathematics and physics. If so, a more precise definition, or an analysis into more primitive concepts, is needed before we can address the problem. Once we have that, I suspect the problem might not look so intractable.

References

1. S. Wenmackers. Children of the cosmos. FQXi Essay Contest Entry. Available at http://fqxi.org/community/forum/topic/2492, 2015.
2. Discussion thread for "Mathematics is Physics". http://fqxi.org/community/forum/topic/2364, 2015.
3. Eugene P. Wigner. The unreasonable effectiveness of mathematics in the natural sciences. *Communications on Pure and Applied Mathematics*, 13(1):1–14, 1960.
4. W. V. Quine. Epistemology naturalized. In *Ontological Relativity and Other Essays*, chapter 3, pages 69–90. Columbia University Press, 1969.
5. David Papineau. Naturalism. In Edward N. Zalta, editor, *The Stanford Encyclopedia of Philosopy*. Spring 2009 edition, 2009. URL: http://plato.stanford.edu/archives/spr2009/entries/naturalism/.
6. Willard Van Orman Quine. Two dogmas of empiricism. *The Philosophical Review*, pages 20–43, 1951.
7. Hilary Putnam. Mathematics without foundations. *The Journal of Philosophy*, 64(1):5–22, 1967. URL: http://www.jstor.org/stable/2024603.
8. Hilary Putnam. Is logic empirical? In Robert S. Cohen and Marx W. Wartofsky, editors, *Boston Studies in the Philosophy of Science*, volume 5, pages 216–241. D. Reidel, 1968.
9. Hilary Putnam. *Philosophy of Logic*. George Allen & Unwin Ltd., 1972.
10. Max Tegmark. *Our Mathematical Universe*. Knopf, 2014.
11. Leon Horsten. Philosophy of mathematics. In Edward N. Zalta, editor, *The Stanford Encyclopedia of Philosophy*. Spring 2015 edition, 2015. URL: http://plato.stanford.edu/archives/spr2015/entries/philosophy-mathematics/.
12. P. Benacerraf. Mathematical truth. In P. Benacerraf and Hilary Putnam, editors, *Philosophy of Mathematics: Selected Readings*, pages 403–420. Cambridge University Press, 2nd edition, 1983.
13. Dimitris V. Kalamaras. SocNetV. v1.5. URL: http://socnetv.sourceforge.net.
14. Rka Albert and Albert-Lszl Barabsi. Statistical mechanics of complex networks. *Reviews of Modern Physics*, 74(1):47–96, 2002. eprint arXiv:cond-mat/0106096. doi:10.1103/RevModPhys.74.47.
15. Elaine Landry and Jean-Pierre Marquis. Categories in context: Historical, foundational, and philosophical. *Philosophica Mathematica*, 13(1):1–43, 2005. doi:10.1093/philmat/nki005.
16. Stefan Banach and Alfred Tarski. Sur la dcomposition des ensembles de points en parties respectivement congruentes. *Fundamenta Mathematicae*, 6:244–277, 1924.

[5]Technically, it grows like the logarithm of the number of nodes.

17. K. Wharton. Mathematics: Intuition's consistency check. FQXi Essay Context Entry. Available at http://fqxi.org/community/forum/topic/2493, 2015.
18. L. Carroll. What the tortoise said to Achilles. Mind, 104(416):691–693, 1895. doi:10.1093/mind/104.416.691.
19. P. D. Grunwald and P. M. B. Vitanyi. Algorithmic information theory. In P. Adriaans and J. van Benthem, editors, *Philosophy of information*, Handbook of the philosophy of science, chapter 3c, pages 281–317. Elsevier, 2008. arXiv:0809.2754, doi:10.1016/B978-0-444-51726-5.50013-3.

My God, It's Full of Clones: Living in a Mathematical Universe

Marc Séguin

Abstract Imagine there's only math—physics is nothing more than mathematics, we are self-aware mathematical substructures, and our physical universe is nothing more than a mathematical structure "seen from the inside": that's what Max Tegmark's *Mathematical Universe Hypothesis* (MUH) proposes. In this paper, I will discuss some consequences of the MUH. While Tegmark claims that the MUH implies the existence of an enormous but *finite* multiverse (to avoid the measure problem that occurs when you try to calculate probabilities within an infinite ensemble), I will argue that it implies the existence of the largest imaginable multiverse, the *Maxiverse*, where every imaginable conscious observation is guaranteed to happen. I will attempt to explain why, of all the worlds in the Maxiverse, we happen to live in one that can be understood by physical laws simple enough to be discovered (or, at least, approximated well enough for predictive and technological purposes). I will explore the issue of personal identity in the context of a Maxiverse that contains an infinite number of exact clones of myself, and discuss the *Maxiverse Immortality Argument*, the idea that I should expect my future subjective experience to be unbounded. Finally, I will consider the question of whether or not the Maxiverse hypothesis can make predictions that can be put to the test.

Math from Nothing and Your Physics for Free

What is the relationship between mathematics, the study of abstract structures, and physics, the study of the physical world? Since the theories of modern physics are formulated in the language of mathematics, we can try to express the relationship between mathematics and physics by the equation

$$Physics = Math + ?$$

M. Séguin (✉)
Physics Department, Collège de Maisonneuve,
Montréal, QC, Canada
e-mail: mseguin@cmaisonneuve.qc.ca

A. Aguirre et al. (eds.), *Trick or Truth?*, The Frontiers Collection,
DOI 10.1007/978-3-319-27495-9_4

where "?" stands for one or more other fundamental ingredients which make up physics.

Leaving aside for a moment the meaning of the "?", we should ask ourselves if the general form of this equation even makes sense. The equation states that mathematics is an ingredient of physics, but for this to be possible, mathematics has to have an independent existence. If you believe, like some physicists do, that mathematics is something invented by mathematicians (and which sometimes happens to be useful to describe the physical world), then it makes no sense to say that it is one of the fundamental ingredients which make up physics. Therefore, the equation only makes sense from the point of view of **mathematical platonism**, the belief that mathematical structures "exist by themselves".

Let us compare a number which is thought about daily by millions of people, like the number 5, and a one billion digit number which has never been thought about by any mathematician or represented in any computer program. According to mathematical platonism, the later truly and independently exists, in exactly the same way that the number 5 exists. Those who object to mathematical platonism consider that we should not use the word "exist" to qualify an abstract entity which has never been "embodied" physically, or, to the very least, thought about by a physical being. But if you insist that something only truly exists if it is physically embodied, then you must explain what physical embodiment is, and what is so special about it. Modern physics reveals that matter is made of fundamental building blocks, essentially electrons and quarks in the case of ordinary matter. Some of the properties that we associate with matter at our scale (like texture and color) are *emergent* properties that do not exist at the level of electrons or quarks, where we encounter "wave-particles" that do not have precise shapes, positions or trajectories. Some properties of matter that we perceive at our scale (like mass and electric charge) do manifest themselves at the level of electrons and quarks, since each of them has a certain mass and a certain charge. But it is not clear what mass and charge are: they seem to be basic attributes of the particles—we can quantify them with a number, but we do not know what they truly are. Quantum field theory makes things even more… fuzzy. The wave-particles are excitations of the fields, very abstract entities that fill all space. Quantum fields carry mass, charge and other properties, but they are so "ethereal" that they might as well be pure mathematical structures.

Therefore, perhaps the "?" in our equation stands for nothing:

$$Physics = Math + nothing\ else$$

This radical proposition is equivalent to the **Mathematical Universe Hypothesis (MUH)** that cosmologist Max Tegmark proposed in his 1998 paper *Is* "the theory of everything" *merely the ultimate ensemble theory?*, and expounded in his 2014 book *Our Mathematical Universe*. For Tegmark, physics is *nothing more* than mathematics, or, more precisely, our physical universe is *nothing more* than a mathematical structure: the patterns and laws which make up the mathematical description of our physical world *are physical reality itself*. According to the MUH, the physicality of our universe can be completely explained by the emergent properties of the underlying abstract mathematical structure. The fundamental level of our reality is made of relationships between entities which are themselves purely abstract: it's "all structure, no stuff"! In his 2012 book *Why Does The World Exist?*, Jim Holt calls this view **cosmic structuralism**.

An interesting parallel can be made between the MUH and the "Physical Life Hypothesis",

$$Biology = Physics\ (Chemistry) + nothing\ else$$

which has replaced (at least, among scientifically minded people) the once popular "Vitalism Hypothesis",

$$Biology = Physics\ (Chemistry) + \text{``}Life\ spark\text{''}$$

Many people intuitively reject the MUH because of a "gut feeling" that mathematical structures and physical structures cannot be equivalent: after all, mathematical structures are abstract, while physical structures are, well, "physical", which means "concrete", "tangible", "material". But if you accept that a living being can be thought of as nothing more than a complex arrangement of atoms obeying the laws of physics, is it really that hard to accept that a physical universe can be thought of as nothing more than a complex mathematical structure?

In his 1988 book *A Brief History of Time*, the physicist Stephen Hawking wrote:

> [A physical theory] is just a set of rules and equations. What is it that breathes fire into the equations and makes a universe for them to describe? The usual approach of science of constructing a mathematical model cannot answer the question of why there should be a universe for the model to describe. Why does the universe go to all the bother of existing?

According to the MUH, you are a complex mathematical substructure embedded in a larger mathematical structure, but the particular way your substructure is arranged makes you self-conscious, and this consciousness allows you to observe the structure "from the inside": what makes our universe "physical" is, fundamentally, the contemplation of its mathematical structure by the self-aware substructures it contains.

But why does our universe exist? If the fundamental level of reality is made of some physical stuff, to make a universe, you need to get the stuff first. Even if it turns out that our universe sprang from a tiny fluctuation in some primordial "false vacuum" quantum field, you still need to get some false vacuum from the store! On the other hand, if the basic level of reality is an abstract mathematical structure, and mathematical structures exist by themselves, in a "timeless" and "eternal" way, there is no longer any issue. So, if you can get over the initial shock and disbelief that physics can be the same thing as mathematics "seen from the inside", cosmic structuralism as expressed by the MUH explains in a simple way why our universe exists. Of course, you get more, *much more*, in the bargain…

What Part of ∃ Don't You Understand? Welcome to the Maxiverse

If the MUH is true, and our universe is nothing more than some complex mathematical structure, there exists an enormous number of other physical universes, one for each mathematical structure which has the right properties to be a physical universe. There are an infinite number of mathematical structures (since natural numbers themselves, one small subset of all possible mathematical structures, are infinite in number). It is likely that most mathematical structures do not have the correct properties to be physical universes, or are not complex or regular enough to contain self-aware substructures that can "feel" them from the inside—but those who do are probably infinite in number. Therefore, the MUH predicts the existence of an enormous multiverse that contains a huge number of physical universes, our universe being one of them. Some of these universes are quite similar to our own, others contain exotic phenomenon which obey completely different laws of physics. Tegmark calls this the **Level IV multiverse** (to distinguish it from lesser multiverses which regroup universes that share some similarities to our own); in his 2011 book *The Hidden Reality*, Brian Greene uses the term **Ultimate Multiverse**. If math is all it takes to have a universe, it makes no sense to believe that some universes exist while others don't. As Greene explains,"There's no switch that turns math "on"." Every universe in the Ultimate Multiverse is as real as our own: what we consider to be our universe is simply the particular mathematical structure that we happen to find ourselves in.

Many physicists have a hard time with any hypothesis which implies the existence of a multiverse. One possible reason is that, if the multiverse exists (and especially the Ultimate Multiverse), the physics of our universe is only one physics among many, which means that physicists in our universe do not discover and ponder the Universal Principles of All of Reality, but merely the local bylaws of our little corner of it. Another popular criticism of any hypothesis which implies a multiverse is saying that it violates Occam's razor, because it postulates an enormous, potentially infinite number of unobservable universes to explain our observable reality. But it all depends on what you try to minimize: the number of things that exist, or the number and complexity of the principles that define their existence. If you want to explain why one or only a few universes exist, you must specify the precise laws they obey and their initial conditions (at least). You must also specify and justify the rules which select these universes to be real while relegating all other possibilities to the dustbin of existence. Specifying the initial conditions alone might necessitate a mind boggling amount of information. On the other hand, to describe completely the Ultimate Multiverse, one short sentence is enough: *the collection of every mathematical structure which has the correct properties to correspond to a physical reality*. Or you could describe the Ultimate Multiverse as Brian Greene does, in only eight words: "Across the multiverse, all math gets its due."

The idea that every possible universe is as real as our own has been proposed by many philosophers before Tegmark's formulation of the MUH. In his 1936 book *The*

Great Chain of Being, historian Arthur Lovejoy coined the term **principle of plenitude** to describe the idea that the universe contains all possible forms of existence: he surveyed the presence of this idea throughout the history of philosophy, tracing it back to Plato's theory of forms in the 4th century BC. Modern theories which explicitly state that every possible universe exists have been formulated by Robert Nozick in his 1981 treatise *Philosophical Explanations* (based on what he calls the **principle of fecundity**) and by David Lewis in his 1986 book *On the Plurality of Worlds* (based on his theory of **modal realism**).

Indeed, for some philosophers, the idea that everything exists has an elegance and simplicity which makes it irresistible. No matter what the ultimate cause of existence is, we know that it has been able to create an actual world at least once, since we observe such a world. What could prevent this cause from acting again to create another world, and another, and another? And even if a given cause eventually "runs out of steam", being an ultimate cause, it exists by itself: if it instantiated itself once, what could prevent it from instantiating itself once more, creating other worlds? How could this process fail to create an infinite number of worlds which encompass all possibilities? Tegmark's explicit formulation of the MUH only makes it less problematic to believe in such an infinity of worlds, because all you need is math, and "math is cheap", existing by itself without using up any "limited natural resources".

It is interesting to note that Tegmark, although he believes in the MUH and in an enormous Level IV multiverse, takes great care to explain that he *does not believe* that every imaginable universe exists:

> The Level IV multiverse can be thought of as a smaller and more rigorously defined reality by virtue of replacing Lewis's "all possible worlds" by "all mathematical structures". The Level IV multiverse does not imply that all imaginable universes exist. We humans can imagine many things that are mathematically undefined and hence don't correspond to mathematical structures. Mathematicians publish papers with existence proofs that demonstrate the mathematical consistency of various mathematical-structure descriptions precisely because to do this is difficult and not possible in all cases (*Our Mathematical Universe*, p. 351)

In his 2008 paper *The Mathematical Universe*, Tegmark explains that although a mathematical structure is made of objects with relations between them, not all theories are mathematical structures, because some "objects", like God in the theory "God created Adam and Eve", may not be definable, even in principle, in a rigorous enough way to serve as objects within a mathematical structure. Tegmark is also very cautious when it comes to infinity, because it makes probabilities within the multiverse virtually impossible to compute in a non-arbitrary way. (I will come back to this issue in section "Why is our world so lawful and simple?".)

My opinions on these issues diverge from those of Tegmark. I do not think that it is possible to imagine a physical world containing self-aware substructures which could not be described by *some sort* of mathematics. Therefore, there is a no real difference, from a practical point of view, between the Level IV multiverse implied by Tegmark's MUH and the "All Imaginable Universes" of the philosophers. I do not think it's possible to imagine an abstract structure which could not, in some way, be described by mathematics: there is no "Level V" multiverse made of non-mathematical structures, because mathematics is the general study of structures. Of course, one can imagine a gigantic, convoluted, ugly, unwieldy and irregular structure which would appear, at first glance, unmathematical. But from the point of view of an infinitely intelligent mathematician, even such a structure would be describable in a mathematical way. Such a position may seem to fly in the face of Gödel's incompleteness theorem, but it is important to realize that the theorem only means that there exist true mathematical statements that can never be proven by a finite set of axioms manipulated by a finite mind. Non-infinitely intelligent mathematicians will never be able to fully capture the whole of mathematics within a consistent axiomatic system, but this would even be the case without Gödel's incompleteness, because mathematics is infinite. Gödel's incompleteness theorem only means that some mathematical truths will forever remain out of reach from any finite mathematician. It does not mean that the MUH is ill-defined, and I do not believe, like Tegmark does, that it could be necessary to restrict the MUH to finite "computable" functions to make it work (he calls this reduced hypothesis **CUH**, for **Computable Universe Hypothesis**).

It should be obvious from the previous paragraph that I also believe that we should fully embrace the concept of infinity. In fact, I think the MUH implies the existence of an *infinite* multiverse, and that many of the universes it contains are themselves infinite. I believe that the multiverse implied by the MUH is the maximal cosmological playground, the biggest possible multiverse, containing every imaginable physical reality, and generating every imaginable conscious observation: I propose to call it the **Maxiverse**. In the rest of this paper, I will discuss some challenges raised by the Maxiverse hypothesis: the difficulty in reconciling the Maxiverse with the fact that we observe a lawful and relatively understandable reality; the problem of finding out which of the infinite mathematical structures consistent with the world we observe is the "actual one"; and the question of the perilous status of a theory which predicts that every imaginable conscious observation is realized somewhere.

Why Is Our World so Lawful and Simple?

In his 1999 book *Robot: Mere Machine to Transcendent Mind*, computer scientist and artificial intelligence pioneer Hans Moravec was one of the first to explore the idea that the Maxiverse might be real. In an interview from that same year reprinted in David Jay Brown's 2005 book *Conversations on the Edge of the Apocalypse*, Moravec clearly explains one of the biggest challenges of the Maxiverse hypothesis:

> So if our world exists platonically, but in a sea of other possibilities, you then have to ask the question: Why is our world so boring? In the space of all possible worlds, there's a world in which in the next second you sprout wings on your head and your nose grows into an elephant's trunk. There are worlds like that that exist in the space of all possible worlds. So why doesn't that really happen to us? Why does our world seem to be so boring, so tied to these simple physical laws that we're only recently starting to elucidate?

Our world is clearly regular: it obeys stable physical laws, and those laws are relatively simple, in the sense that we can at least approximate them in such a precise way that we can predict the evolution of physical systems and build technological contraptions which exploit that knowledge. For instance, we can plan years in advance for something as complex as a robot rover mission to Mars, and carry out the plan with success.

In the cosmological smorgasbord of the Maxiverse, where every imaginable universe exists, we should expect baroque, irregular and chaotic worlds to greatly outnumber lawful and predictable worlds such as ours. Our type of universe would then be highly unlikely, which would make the Maxiverse hypothesis highly suspect. One way out of this dilemma is to suppose that some worlds in the Maxiverse are more probable than others, and that regular worlds like ours have a higher *measure* (probability) than irregular worlds where you sprout wings on your head.

Now, if the Maxiverse, as I believe, is an infinite ensemble of universes, the very notion of what is likely or unlikely becomes ill-defined. In a finite ensemble, it is straightforward to evaluate the fraction of its members which have a given property: for instance, if you have a bag which contains a finite number of black or white marbles, the question "What fraction of balls is white?" has a definite answer, because you can sort the balls in two piles and count them. On the other hand, in an infinite ensemble, the question "What fraction of the members of the ensemble has some property X?" is not well defined. In his book (p. 313), Tegmark takes the example of the infinite set of all positive integers (1, 2, 3, 4, 5, 6...). It seems obvious that half its elements are even: if you analyze any portion of the list, you observe that odd and even numbers alternate, and you naturally conclude that half of the numbers are even. The problem is that, *because the set is infinite*, you can imagine other systematic orderings which do not lead to the same conclusion. For instance, if you make a list by starting with 1, 2, 4 and extend it by always adding the next larger *odd number* followed by the next two larger <u>even numbers</u>, you get the sequence *1*, <u>2</u>, <u>4</u>, *3*, <u>6</u>, <u>8</u>, *5*, <u>10</u>, <u>12</u>, *7*, <u>14</u>, <u>16</u>, *9*, <u>18</u>, <u>20</u>... This infinite list is complete, because no positive integer is left out. But now, if you try to evaluate the fraction of even numbers by analyzing some portion of the list, you conclude that *two thirds* of the numbers are even! This example seems contrived, because it seems "obvious" that the order 1, 2, 3, 4, 5, 6... is the most natural one. But if you try to list universes in a set that contains an infinite number of them, there will not be any obvious, natural and unique way to order them, and it will be impossible to unambiguously calculate the fraction of these universes associated with a given property. This ambiguity concerning probabilities within infinite sets is known as the **measure problem**, and it is the main reason why Tegmark hopes that his MUH (or CUH) can be reigned back so it implies only a finite set of finite universes.

I believe that the way out of the dilemma involves something similar to the **anthropic principle**, the somewhat tautological statement that it's impossible to observe a universe whose properties are incompatible with life, since we couldn't exist there in the first place—therefore, the properties of our universe must be compatible with the existence of life. In the same way, we observe that our universe stays lawful and predictable, even if there are many scenarios where it doesn't, because in these scenarios, our consciousness immediately ceases to exist—so we never observe them. In the above mentioned interview, Moravec explains:

> Your neurons wouldn't work if chemistry changed in some way. Your consciousness is an interpretation of the way your neurons work, and your connection with the world. A lot of your experiences depend on everything working just the way it does. If the speed of light were to change, certain chemical reactions would alter, and your consciousness would probably be gone. But pretty much if the laws of physics were altered in any way, your consciousness would no longer work the way that it does. In those other worlds, if that's all that happened, you would no longer exist. So you can't find yourself in those worlds. Maybe some other things could change that bring back your consciousness, but that would be like another coincidence that would have to happen. The odds of that happening are small. So the most likely world that you will find yourself in the next moment is one that's just a continuation of the world that you're in right now, because nothing has to change. The mechanism that you have all this investment in—this evolutionary and biological growth investment—just continues. The only other question is: Why are we in this kind of world in the first place? And again, now that we're in it, we're kind of stuck. Probably it is the case that this is the simplest world, the world that required the least number of coincidental starting positions to produce us.

If our universe was any less lawful and relatively understandable than it is, the ideas explored in this article would probably never have been considered, and a forum like the Foundational Questions Institute would not exist. Could the ultimate answer as to why our universe exhibits such lawfulness and relative simplicity be the fact that it contains philosophers and scientists such as Robert Nozick, David Lewis, Hans Moravec, Stephen Hawking, Brian Greene and Max Tegmark?

Lost in the Maxiverse

Where am I in the Maxiverse? According to the MUH, my consciousness, my sense impressions and my memories are nothing more than a complex but finite self-aware substructure. In the infinite ensemble of all possible mathematical structures, there exists an infinite number of exact copies of this finite substructure. Each of these "clones" has exactly the same experiences that I have, and believes itself to be me. Each clone is embedded in a different larger mathematical structure, in a different "physical world". But these differences do not have any practical impact on my clones, since they are all identical to me. (There exists also an infinite number of slight variations of my substructure, and an infinite number of slightly more different variations, ad infinitum, but let us concentrate on my exact copies.) Some of my clones are part of mathematical structures which correspond to 13.8 billion-year-old infinite universes embedded in finite post-inflationary bubbles (as described by the eternal inflation theory of Alan Guth and Andrei Linde): details of these universes that do not influence the clones (like what happens outside the cosmic horizon in each universe) differ from universe to universe. I have some clones that are part of mathematical structures which correspond to powerful computer simulations run by post-humans histo-rians to better understand a crucial part of human history, the pre-singularity decades at the beginning of the 21st century. I also have clones that are part of mathematical structures which correspond to the playgrounds of mad transdimensional superintelligences, where they play the role of existential pets.

In a deep, fundamental sense, I believe that *all these clones are me*: I live simultaneously in an infinite number of larger contexts which differ only in unobservable ways. From one perspective, each of my clones is embedded in a different larger structure, which should make it possible to distinguish between them. But from another perspective, my self-aware substructure completely defines me (because it includes my consciousness and my sense impressions), and this substructure is one unique mathematical structure. There is no way to tell which perspective is the correct one: my clones are one unique mathematical structure, that occupies "one spot" in the Maxiverse, but at the same time, they are scattered in an infinite number of copies throughout the Maxiverse. In the non-space of all mathematical structures, it is impossible to self-locate: we are, each and every one of us, fundamentally lost in the Maxiverse.

BOBBY HIDDY, WIKIPEDIA

If you have trouble with the idea that you are living simultaneously in an infinite number of different but indistinguishable contexts, the analogy of a stretch of road that carries two different road numbers might help. In Quebec, highway 20 east and highway 55 north share the same stretch of road for a few kilometers. Suppose you wake up in the passenger seat on that stretch of road, with no memory of getting into the car. It makes no sense to argue about which road you *really* are on. Of course, once you get to the exchange when the two roads go their separate ways, you will wind up on one and only one road. Suppose the driver continues on highway 55: you could conclude that you have been on highway 55 since you woke up, *but only in retrospect.* In the same way, if you consider all the ways that your self-aware substructure can be embedded in a larger mathematical structure, you exist right now in a 13.8 billion-year-old universe, on a rock and metal and water planet where it is 2015 on the local calendar, but also inside a powerful computer running an historical simulation in a remote solar system where the planets have been dismantled and converted into computronium. And there are many, many other scenarios that are equally true of your current situation. It makes no sense to ask which is the correct one, even though, if you are lucky, you might eventually be able to eliminate some possibilities, *in retrospect.*

Life and Death and Life in the Maxiverse

I go to sleep on the night of March 4, 2015, right after submitting my FQXi essay. What can I expect tomorrow morning? If the MUH is true and implies the existence of the Maxiverse, there are many, many possible answers. In some universes, my self-aware substructure has been terminated during the night by an unlucky meteor strike (a small meteorite, sufficient to break through a roof and kill a person in his sleep, would not be detected beforehand). In other universes, a wave of decaying vacuum travelling at the speed of light has instantaneously disintegrated the Earth while I slept, taking humanity by surprise and painlessly wiping it out. In another universe, I was part of an historical simulation that got axed because of budget cuts, and I have been terminated and erased during the night. Of course, I cannot ever become aware of these possibilities: I can only be aware of the scenarios where I continue to exist.

Tomorrow morning (from my perspective), I can be any of the self-aware substructures in the Maxiverse that remember going to sleep as me: let's call them my **F-clones** (F for "future"). Some of my F-clones are very surprised to discover that something drastic has happened during the night. Some find themselves in a strange setting, with an error message floating in mid-air explaining that their historical simulation is being shut down, but that they will be taken care of according to the ethical rules of the posthuman research institute that ran the simulation (budget permitting, of course). Some of my F-clones find themselves in a mathematical structure where the Christian Last Judgment has begun during the night, and they stand in line at the Pearly Gates in freshly tailored white and gold robes (or are they blue and black?), waiting to be processed. But of course, I know (based on my previous experiences) that these unusual scenarios are unlikely: I expect to wake up in my bed and lead a more or less ordinary day, which must indicate that somehow (despite the measure problem), my F-clones which correspond to these ordinary scenarios greatly outnumber the other ones.

As long as waking up normally tomorrow morning has a reasonable probability of happening, I should expect that my future subjective experience will remain bound to the lawful and regular universe that I have gotten to know. But if I'm very old or terminally ill, at some point, some other category of scenarios will become more probable. In his 1998 essay *Simulation, Consciousness, Existence* (which is identical to the last chapter of his book *Robot*), Moravec describes how, in the space of all possible worlds, our subjective experience can always find a way to continue (an idea I will refer to as the **Maxiverse Immortality Argument** in what follows):

> Our consciousness now finds itself dependent on the operation of trillions of cells tuned exquisitely to the physical laws into which we evolved. It continues from moment to moment most simply if those laws continue to operate as they have in the past. Thus, with overwhelming probability, we find the laws are stable. In the space of all possible universes, we are bound to the same old one. As long as we remain alive.
>
> When we die, the rules surely change. As our brains and bodies cease to function in the normal way, it takes greater and greater contrivances and coincidences to explain continuing consciousness by their operation. We lose our ties to physical reality, but, in the space of

> all possible worlds, that cannot be the end. Our consciousness continues to exist in some of those, and we will always find ourselves in worlds where we exist and never in ones where we don't. The nature of the next simplest world that can host us, after we abandon physical law, I cannot guess. Does physical reality simply loosen just enough to allow our consciousness to continue? Do we find ourselves in a new body, or no body? It probably depends more on the details of our own consciousness than did the original physical life. Perhaps we are most likely to find ourselves reconstituted in the minds of superintelligent successors, or perhaps in dreamlike worlds (or AI programs) where psychological rather than physical rules dominate.

Because he doesn't think that the MUH implies an infinite Maxiverse, but only a finite Level IV multi-verse (to avoid the measure problem), Tegmark has doubts about the soundness of this immortality argument. Nevertheless, on page 220 of *Our Mathematical Universe*, he writes:

> But who really knows? When one fateful day in the future, you think that your own life is about to end, remember this and don't say to yourself *There's nothing left now*—because there might be. You might be about to discover firsthand that parallel universes really do exist.

What Is the Maxiverse Good For?

In the preceding sections, I have tried to build a case for the MUH and the Maxiverse hypothesis. Most days, I believe that the Maxiverse hypothesis could be the ultimate answer to the question of Life, the Universe and Everything. But I am well aware that the Maxiverse hypothesis will be considered, by many, as a meaningless pipe dream that only someone who has lost contact with reality could entertain. Indeed, there are many reasons to reject the Maxiverse hypothesis. One major reason is the unfortunate fact that it cannot make any predictions that can be tested, since, in the Maxiverse, any imaginable observation occurs somewhere. Consider the most bizarre, seemingly illogical and far-fetched observation that you can think of. Because parts of the Maxiverse correspond to simulated, "fake" worlds that have been designed by twisted programmers with a weird sense of humor, no observation, no matter how crazy, could contradict the Maxiverse hypothesis. Tegmark claims that his Level IV multiverse is potentially falsifiable, and attempts (not very successfully in my opinion) to defend this position in his book. But we can be sure that somewhere in the Level IV multiverse, in a postmodern replay of the story of Job in the Bible, there exists a playful superintelligence that runs a simulated world containing a clone of Tegmark, feeding him with nonsensical observations specifically designed to shake his faith in the Level IV multiverse and the MUH. It would be fascinating to see how this poor clone responds to the challenge!

If observations cannot falsify the Maxiverse hypothesis, one can hope at least that some observations could strengthen its likelihood. For instance, if after your death in this universe, you find yourself in some afterlife, you could interpret this as a confirmation of the Maxiverse Immortality Argument presented in section "Life and death and life in the Maxiverse". Of course, there would be other ways to interpret any afterlife, and in your new "afterworld", all the fundamental questions that can be asked in this universe would still be there: why does the afterworld exist? Why are the laws of the afterworld the way they are and not in some other way? (You could once again posit the existence of the Maxiverse as an answer!) Moreover, it has been pointed out, by Tegmark and others (for instance, Russell Standish in his 2006 book *Theory of Nothing*) that the Maxiverse Immortality Argument could fail in some cases: if, instead of an abrupt transition between life and death in this universe, you undergo a gradual loss of mental faculties (for example, because of Alzheimer's disease), you may reach a point, before your death in this universe, where you no longer have any memories of the life you just lived. In this case, there is no F-clone in all the Maxiverse that can meaningfully carry your subjective experience into the future. If you don't even remember your life while still alive, what kind of meaningful afterlife could you possibly have?

Another argument against the Maxiverse hypothesis (in fact, against any theory which incorporates seriously the notion of a multiverse) is the belief that it critically undermines the future of theoretical physics. The danger, of course, is to use the Maxiverse hypothesis as an easy way out whenever a given property of our universe seems too difficult to explain. For instance, the measured value of the cosmological constant is much, much smaller (by roughly 120 orders of magnitude) than the theoretically predicted "most probable" value. Some cosmologists seriously consider that the way out of this dilemma is to invoke the anthropic principle applied to a multiverse where universes with every value of the cosmological constant exist: only in universes where the cosmological constant is much, much smaller than the expected value would the conditions be suitable for the emergence of life. But other cosmologists consider that it is too soon to suppose that we will never discover some new physics which will explain naturally the observed value of the cosmological constant. I agree that it is not good scientific practice to appeal to the anthropic principle by default, but it doesn't mean that I think we shouldn't entertain the idea that the Maxiverse can be real—in the same way that I believe we should not build nuclear weapons, but it doesn't mean that I think there shouldn't be any nuclear physicists.

The Maxiverse hypothesis, although not falsifiable in the traditional sense, can motivate some legitimate scientific research. Since the measure problem is such a nuisance in any theory which postulates an infinite reality, more fundamental research at the intersection of probability theory and the study of infinity would obviously be helpful. If the MUH is correct, everything is a mathematical structure, including our minds, and it is our self-awareness and our observations of the world that give it its "physicality": without conscious observers, it wouldn't mean anything to say that some mathematical structures correspond to physical universes, because there would be no one to "feel" the "potential physicality" of these abstract structures. Therefore, the more we learn about the fundamental nature of consciousness, the better we will be able to understand the deep relationships between mind, matter and mathematics.

A completely different criticism of the Maxiverse hypothesis comes from the domain of human values. The mathematical structures which make up the Maxiverse exist by themselves, in a "timeless" and "eternal" way. Therefore, when we strive to make something good in our universe (raising a child, inventing a new medicine, discovering a new theory), we do not change the Maxiverse in any way, we merely "visit" preexisting mathematical structures that have always been part of the Maxiverse. Moreover, every time we succeed in our universe, we can be certain that some of our clones experienced failure in theirs. Does this mean that we should stop trying and stop caring, since all outcomes exist anyway in the Maxiverse? If we believe that the Maxiverse is real, how should we live our lives? The same kind of existential question can be raised concerning the issue of free will, given the fact that, from the point of view of 4-dimensional space-time, the past, present and future of our lives correspond to completed, fully determined 4-dimensional braids. Free will makes no sense from such a perspective, but from *our* perspective of 3-dimensional beings experiencing time in a sequential fashion, free will does have meaning. In the same way, in the context of the Maxiverse, your actions have no impact overall. But from your point of view, limited to one universe, your actions do matter, and you should care. In fact, free will acquires *more* meaning within a Maxiverse, since your inability to self-locate within it means that, essentially, there is no way to predict what you will do next. Some of your clones will do one thing, some will do something else, but since before the fact *you are all of these clones simultaneously*, no one, even an omniscient intelligence, can predict in advance what a particular you will do and experience in a particular universe.

If all this makes your head spin, you're not alone. If you no longer know what to think about the MUH and the Maxiverse hypothesis, you might consider the position defended by Piet Hut in the 2006 article he co-authored with Mark Alford and Max Tegmark, *On Math, Matter and Mind*. Hut argues that we know way too little about mathematics, physics and consciousness to be able to have a coherent discussion about their ultimate nature and relationship. He believes that there are ultimate answers, and that humanity's knowledge might one day be sufficient to successfully tackle these issues: it's just that, at our current level of scientific and philosophical sophistication, we're just not worthy of such deep questions—yet. Maybe he's right, but this will not stop speculative cosmologists from arguing about these deep questions anyway, because that's what speculative cosmologists do. As Steven Weinberg wrote in his 1992 book *Dreams of a Final Theory*, there is no point in "trying to be logical about a question that is not really susceptible to logical argument: the question of what should or should not engage our sense of wonder."

In the end, I believe in the Maxiverse because it is the ultimate playground for thc curious mind. Living forever… across wildly divergent realities… who could ask, literally, for anything more than the Maxiverse? And if I'm right, somewhere within its infinitely complex simplicity, one of my F-clones is having a drink with one of your F-clones, and we're having a big laugh about it all. Cheers!

References

1. Tegmark, Max. “Is “the theory of everything” merely the ultimate ensemble theory?”, Annals of Physics 270, 1–51, 1998, http://arxiv.org/pdf/gr-qc/9704009v2.pdf.
2. Tegmark, Max. Our Mathematical Universe, New York: Alfred A. Knopf, 2014.
3. Holt, Jim. Why Does The World Exist? London: W.W. Norton, 2012.
4. Hawking, Stephen. A Brief History of Time. London: Bantam Books, 1988.
5. Greene, Brian. The Hidden Reality. New York: Knopf, 2011.
6. Lovejoy, Arthur O. The Great Chain of Being. Cambridge, Massachusetts: Harvard University Press, 1936.
7. Nozick, Robert. Philosophical Explanations. Cambridge, Massachusetts: Harvard University Press, 1981.
8. Lewis, David. On the Plurality of Worlds. Blackwell, Oxford, 1986.
9. Tegmark, Max. “The Mathematical Universe”, Foundations of Physics 38, 101–150, 2008, http://arxiv.org/pdf/0704.0646v2.pdf.
10. Moravec, Hans. Robot: Mere Machine to Transcendent Mind. Oxford: Oxford University Press, 1999.
11. Brown, David Jay. Conversations on the Edge of the Apocalypse. New York: Palgrave Macmillan, 2005.
12. Moravec, Hans. “Simulation, Consciousness, Existence”, Carnegie Mellon University, 1998, http://www.frc.ri.cmu.edu/~hpm/project.archive/general.articles/1998/SimConEx.98.html.
13. Standish, Russell K. Theory of Nothing. Charleston, S.C.: BookSurge, 2006.
14. Hut, Piet, Mark Alford and Max Tegmark. “On Math, Matter and Mind”, Foundations of Physics 36, 765–794, 2006, http://arxiv.org/pdf/physics/0510188v2.pdf.
15. Weinberg, Steven L. Dreams of a Final Theory. New York: Pantheon, 1992.

Let's Consider Two Spherical Chickens

Tommaso Bolognesi

4:45 A.M. (Jan 2015, Pisa)

Phone rings. Only one person can call me this early. "Can you make it to be here in the office in 15 min?" boss says, with no preambles. The only admitted answer in these cases is "Yes Sir!" and is implicitly assumed. "Its about three suspected murder cases, and cor blimey if they are not related! We only have to find how. By the way, we are talking 10 thousand bucks… Hurry up, son!" Click.

The strength of our Investigation Agency is, famously, the rapidity of our arrival to the crime scene, wherever it is located in space, time, or… elsewhere. Well, this morning we have really outperformed ourselves!

7:00 A.M. (523 B.C., Crotone, Southern Italy)

Glauco, the young fisherman, is shocked. He can't stop looking at the pale face of his older brother Daippo, whose corpse lies on the sand in front of their humble thatched hut. That's where he found the poor man earlier this morning, back from nightly fishing. As two compassionate women take care of the body, I convince Glauco to answer some questions.

He has no doubts: not an accidental drowning, but homicide.

"Last year Daippo had been admitted to the *mathematikoi*, the restricted group of followers of Master Pythagoras. They are severely required not to disclose their findings, but he could not keep secrets with me. So, one evening he came back home with this device." He shows me a sort of rudimentary, one-string guitar. Must be the famous Pythagorean monochord!

T. Bolognesi (✉)
CNR—Istituto ISTI, Via Moruzzi 1, 56124 Pisa, Italy
e-mail: t.bolognesi@isti.cnr.it

A. Aguirre et al. (eds.), *Trick or Truth?*, The Frontiers Collection,
DOI 10.1007/978-3-319-27495-9_5

"It seems (but we are not supposed to know!) that Phytagoras has been experimenting a lot with this object lately. Nice pleasant sequences of sounds can be obtained by placing a nut under the rope, but only at precise locations, in order to reduce its length by simple fractions—in particular 1/2, 1/3, 1/4, 1/5."

The four locations are indeed marked on the instrument, and now Glauco is plucking the rope, each time positioning the nut at one of them, in sequence. To my surprise, the ascending four-note melody turns out to be *exactly* the opening of Richard Strauss' epic theme 'Also sprach Zarathustra', made famous by the movie '2001, A Space Odyssey'.

"Nice, isnt it?" he continues. "Daippo was fascinated too by the Pythagorean jingle (that's how they call it at the School). But he thought he could do better, and built that thing."

He points to a table where five parallel ropes of equal thickness run a couple of inches above the wooden surface, fixed at a bridge at one side, and pulled at the other, through pulleys, by sand-filled bags of different weights. Basically, a gigantic, fretless bass guitar!

"Following the intuition of his Master, Glauco started by filling the bags with amounts of sand in the same simple ratios: 1 cup in bag 1, 2 in bag 2, up to 5 in bag 5. But the ascending tune obtained by plucking the ropes in sequence turned out to be ugly. Glauco then noticed that the nice interval perceived between the nut-free sound and the sound with the nut at position 1/2 in Pythagoras instrument is the same that we perceive between string 1 and string 4 of his instrument. Essentially, ratio 2 in the first case corresponds to ratio 4 in the second!

This has suggested my brother to *square up* the amounts of sand in the successive bags, filling them with, respectively, 1, 4, 9, 16, 25 sand cups. That's how they're filled now. Oh, sorry, this 'squaring' operation is a technicality that turns out to be useful also in a theorem that Pythagoras… Well, never mind. The end of the story is that, when plucking the ropes of bags 4, 9, 16, 25, in that order, you get back, guess what?

"The Pythagorean Jingle?" I suggest.

"Bingo! Glauco was so excited about it that he started playing the tune over and over. At some point I was so fed up that I decided to go fishing. During the night some member of the brotherhood must have overheard the tune, immediately reporting back to Pythagoras. The end of the story is under your eyes now". He turns to the white cloth wrapped around his beloved brother and explodes in a desperate crying. It is clear that his testimony is over.

I walk to the beach, pick up a thin stick and write on the foreshore Mersenne's formula for the frequency f of a string, as I recollect it from high-school:

$$f = \frac{1}{2L}\sqrt{\frac{sandCups}{\mu}} \tag{1}$$

where L is the rope length, $sandCups$ is the amount of sand in the bag, and μ is the string mass per unit length. The formula tells clearly that one obtains the same

Fig. 1 Two almost spherical, almost self-aware substructures (SAS).

frequency interval by multiplying the string length L by $1/n$, or the sand quantity by n^2. Good, I think I have some useful clues here.

9:00 A.M. (Late June 2008, University of Vermont at Burlington)

A four-bed room in the undergraduate dorm. Four young girls in tears. A boy and two ropes (again!) on the floor. The boy is from Peru', his name is Quipo.[1] As an energetic nurse is attempting cardiopulmonary resuscitation, I interrogate the girls. They are here for the 'π Summer School', a two-week mathematics enrichment programme for teen-agers. Yesterday they had a very exciting lesson on Knot Theory by the distinguished mathematician John Horton Conway, who invited them in front of the young audience to volunteer for an experiment with two ropes. The girls had to stand at the corners of an ideal square, each holding firmly one rope end, and to move in a sort of dance that admits only two kinds of collective moves:

Swap The girls at the upper-right and lower-right corners swap their positions, the former holding the rope up over the latter while moving.

Rotates The whole quartet rotates counterclockwise by 90°.

A fifth volunteer is keeping track of what happens by writing on the blackboard a sequence of numbers. Initially the two ropes are parallel and untangled, one at the upper and one at the lower square edge, and the first number is $x = 0$. At each move a new number x' is added to the sequence: $x' = x + 1$ for *swap*, $x' = -1/x$ for *rotate*. One of the girls hands me a crinkly sheet:

$$tangleList = \{swap, swap, rotate, swap, swap, swap, rotate, swap, swap, swap\};$$

"This is the 10-move sequence we started with", she says.

[1] For the origin of this name, see http://en.wikipedia.org/wiki/Quipu.

"And the result was a real spaghetti tangle", adds a second girl.
"Then Prof. Conway invited us to disentangle it", says the third.
"And everybody in the audience started shouting!", adds the fourth.
"But eventually we made it, and this is the 13-move sequence that completely disentangled the spaghetti", says the first girl, handing me another sheet.

$$untangleList = \{rotate, swap, rotate, swap, swap, rotate, swap, swap, swap, rotate, swap, swap, swap\};$$

"Ok. That was yesterday morning. How about last night?" I ask.
"Well, Quipo had missed the morning lesson, so after dinner we invited him in our room for repeating the experiment." (Blushing.) "We even wrote a simple program to make double-sure that the numeric sequence would bring us back to 0." She shows me her tablet (Appendix 1.)
"So what went wrong?" I ask, after checking that the second list actually terminates with '0'.
"Nothing went wrong. The mathematics was correct!" she proudly declares.
"I see that! But what happened to Quipo?". I'm losing patience.
"Well, for him to better understand the experiment, we invited Quipo to stand… at the center of the square. Then we moved around him, reproducing the two sequences with extreme care. At the end we started pulling, and pulling, and pulling, and pulling... If $x = 0$ the ropes must untangle, isn't it? It's a theorem! It's true! It must be true! Damn!". In saying this she starts crying, and in a moment all four are crying in unison.

Fortunately, they are soon interrupted by the arrival of the gurney. The nurse announces that Quipo might survive, and quickly leaves the room with him, followed by the girls.

I wonder how they select the students for these programmes! No doubt the math correctly describes the behaviour of the physical system of *two* ropes. But, even ignoring Knot Theory, it is clear that, topologically speaking, Quipo was the *third* rope. Different system, different rules! Anyway, I start to glimpse a pattern here.

11:00 A.M. (Circa 2075, New York State)

As I said, we land *anywhere*, fiction included.

I've just entered the huge billiard room. Professor James Priss, two Nobel prizes for physics, is still wearing the dark protective glasses that were handed out to all participants. Edward Bloom, the multi-millionaire who built his fortune by engineering Priss's theoretical findings, is lying on the floor; a hole the size of a billiard ball has crossed his chest, completely removing his heart. All the media circus, gathered here to cover the match between the two life-long friends and rivals, is slowly leaving the room. The event, organised by Bloom for illustrating the exploitability of Priss' Two-Field Theory, has terminated in the most unexpected, tragic way.

The scene is exactly as I pictured while reading 'The Billiard Ball' by Isaac Asimov. The two poles of a huge anti-gravitational device are positioned above and below the billiard table, perfectly aligned with a 12-in hole at the center. Only a few minutes ago everybody could see the glowing cylinder of the zero-gravity field materialize between the two poles, across the hole. Then Bloom invited Priss to take the cue and direct the ball into the cylinder, to be the first in history to experience the effects of the zero-gravity field on a massive body. Priss accepted with apparent reluctance, but he took a significant amount of time, as if pondering behind his dark glasses (which he touched and adjusted several times), before deciding about the optimal trajectory. In light of the life-long rivalry between the two characters, and of Priss' exceptional skills both in physics and in billiard, the suspect of a perfect homicide is quite legitimate.

"Professor Priss, do you think that, once programmed with all the equations of your Two-Field Theory, and all other known theories, including, say, the physics of elastic collisions, a *gigantic supercomputer* could, very hypothetically speaking, ehm... figure out the relation between the incoming and outgoing trajectories and velocities of a billiard ball crossing a zero-gravity cylinder?"

"I suppose your *gigantic supercomputer*", he smiles somewhat sardonically, "is also aware of the exact initial conditions—distribution of planets, stars and so on, around the cylinder, right?" he remarks.

"Oh, sure!"

He takes a deep breath. "My friend, you don't need to beat around the bush! You see, whatever I tell you now will not appear in Asimov's novel, which is the only place where I exist. My worldline is completely determined."

He takes off his dark glasses and hands them to me. "A gigantic computer? That's a century old technology. Ever heard about Google Glass v.18? You see, it all fits in the side temples. Look."

As I wear the glasses, an augmented reality scene appears to me, with stylised billiard balls and virtual dotted trajectories dancing before my eyes.

"What not even Asimov knew is that a mole in Bloom's team was keeping me informed about his plans. So I had plenty of time for programming this little toy. At the right moment, it was easy to swap glasses and have them assist me in the perfect shot."

1:00 P.M. (Jan. 2015, Between New York and Pisa)

I'm driving back to the Agency headquarters, in good company with the 'Police' and a triple sandwich. The illumination comes as Sting starts one of my favourites: 'Murder by Numbers'.

That's obvious! The common denominator of the three crimes is *the unreasonable correspondence between mathematics and reality*. We reliably use simple numeric ratios for describing nice musical intervals, algebraic operations for describing and predicting the behaviour of knots, and compact equations for telling the behavior of vibrating strings, massive bodies and fields. I think I have a culprit, and I can't wait calling the boss.

"So, your point is that some people here have died because math supposedly describes the physical world fully, reliably and persistently?"

"Exactly! Imagine a world *not* ruled by mathematics: the Pythagorean jingle could't have been conceived, let alone reproduced, and the sequence of knots, or the trajectory of the billiard ball, could not have been directed to their lethal final states. Can you imagine such a messy world?"

"Of course I can!", says the boss "That's exactly *our* world! I can mention numberless cases of people who were killed precisely because the world is *not* ruled by math. Randomness is the greatest killer! Do you remember that case down in Vegas? Was it not the outcome of a totally random event?".

"You mean the Russian roulette case? Well, the position where the revolver cylinder stopped was *not* random. Of course it was unpredictable to the poor guy but, had he been given all the initial values, he could have computed the outcome."

"Look, if math were so effective in describing what *really* happens in this world, I would have hired a mathematician, not you! Math may well be *unreasonably* effective, but only for describing a few *unreasonably* idealized cases! Remember the brilliant brains we consulted for the henhouse case? They started with *two perfectly spherical chicken*! (Fig. 1). Adding a third immediately got them into trouble! In the Russian roulette case, initial conditions involve the movement of the hand which transmits the rotation, the brain that controlled that hand, the neurochemical processes in the brain, the physics behind them, and so on ad infinitum. How can you expect to mathematically tame this infinite mess?"

"Not ad infinitum. Maybe there is a bottom" I suggest.

"Then you must fully specify the initial conditions down there—real variables that require infinite decimal expansions: unfeasible. Add the butterfly effect, and things become *totally* unpredictable. Randomness strikes back!"

"Maybe the bottom is discrete and finite instead, and can be described *completely*, like the input data of a computer program" I timidly propose.

"Ok buddy, I admit that we have often cracked hard cases by making bold hypotheses, but this is the craziest conjecture I've ever heard!" Click.

That made me thirsty. As I grab the bottle with one hand, driving with the other, a voice from the back makes me jump. "The right question is not *whether* mathematics describes the physical world—which is obviously true—but *why*!"

"Prof. Priss, what are you doing here?"

Priss quietly smiles at me. "I'm just taking advantage of my status. But let's not digress. The topic is intriguing, and I hear you hire scientists for consultancy. Two Nobel prizes are not peanuts, and Physics must have progressed a bit from 2015 to 2075, don't you think?"

"Of course! You mean that by 2075 everything has been figured out?"

"Unfortunately not. But first, how about a pizza?"

2:00 P.M. (Jan 2015, Pizza Place Near Pisa)

After a 'napoletana' and two glasses of local wine, Priss is ready for conversation. The following is a faithful transcript from my voice recorder.

"So let's start where your phone call was interrupted. You were completely right: physical reality *does* sit on a discrete, *deterministic* bottom layer. This is clear now—I mean, in 2075! It was 't Hooft who gave the strongest contribution to the grand return of classicality, and the downgrading of quantum mechanics to the status of a mere tool. In his 2014 paper [6] he started showing that Einstein was right: *at its most basic level, there is no randomness in nature, no fundamentally statistical aspect to the laws of evolution.*"

Now, more than *sitting* on that mathematical layer, physical reality *is* that layer, and that layer is all there is: all the rest is redundant baggage—convenient names and concepts that we humans have invented for structuring our descriptions of that reality, for matching the limited bandwidth of our senses and lab equipment. Understanding physical reality means simply *discovering* the mathematical properties of that glimmering piece of math at the bottom."

"Wait! Are you saying that those bizarre ideas attributing an equal status of existence to *all* mathematical structures—the platonic Level IV multiverse—are still being considered in 2075?" [7, 8]

"There is nothing bizarre in viewing the fabric of the universe as a nice piece of math. What else? Take any slice of the spacetime cake and try to provide a baggage-free description of it. When *human baggage allowance* drops to zero (see, I'm using Tegmark's terminology) what is left can only be a mathematical structure. *What else*?"

"Put it differently", he continues, "as you increasingly magnify the fabric of reality, familiar properties tend to vanish. The smell of this pizza? The taste of this wine? The mass of that particle? All gone. What else can be left, other than a purely abstract structure? Now, what is the most abstract, featureless thing that you know? One without smell, color, extension, spin?"

"The point!"

"Good. You start to follow! So now take a few points and some relations among them. What mathematical object do you get?"

"A graph?"

"Fair. What properties does it have?"

"None! No smell, no taste. Nothing!" I am really starting to follow.

"Wrong! It has symmetries, and they can offer you a lot."

"Oh, ok. Anyway, a graph is only one out of many possible abstract structures. How about the others? Are there infinite parallel mathematical universes?"

"Few physicists still worry about this question, in 2075."

"Why?"

"Because of the Priss-Gödel-Priss Theorem (2031):

> All mathematical structures entailing conscious entities: (i) are defined by total, computable functions, and (ii) are isomorphic.

The first part proves that a physical universe hosting consciousness only requires *total computable functions*, those that can be computed by a Turing machine that halts for each input. The Theorem proves and refines a conjecture by Tegmark [7, 8], and is consistent with a fundamental discreteness assumption, since the computable functions from *N* to *N* (*N* are the naturals) can be put in one-to-one correspondence with *N* itself. As you know, the set of all functions from *N* to *N* is much larger than that; those that are left out are not computable; it is remarkable that we don't need all that stuff for defining conscious entities. On the other hand, the ontological status of mathematical structures involving non-computable functions, or even restricted to computable functions, but not entailing consciousness, is, in my opinion, irrelevant". Priss takes a sip of coffee, and continues.

"Well, the story behind the proof is also quite interesting. Initially I started with simple toy physical universes—no life, no consciousness, say just flat Minkowski space—trying to associate them with the right mathematical structures. Easy, you might guess. No way! As I was kicking a non-computable function out of the door, another was entering through the window! Then I went the opposite extreme, and considered a system of two moderately conscious entities having an interaction in empty space—what got to be known as Priss' *two-chickens problem* (Fig. 1)—modelling them along the lines of Tononi's seminal work on consciousness and integrated information [1]. Surprisingly, all non-computable functions disappeared for good, and all the other pieces of the inert physical world popped out and found their place in the picture. *In a way, it is the very presence of consciousness that makes the physical world so simple, and mathematically describable*."

(Mental note: this is definitely going to be the grand opening of my report to the boss.)

"The second part—the isomorphism result—essentially means that there is *only one* possible structure which can host conscious beings like you and me… well, maybe just you. Of course, the Theorem does not put any limitation to the variety of shapes that consciousness may assume in our unique universe, which are not restricted to soliton forms. A generalisation of the two-chicken system, for example, led me to discover sophisticated forms of consciousness in vacua from the Standard Model—something that popular wisdom has believed since antiquity. Needless to say, symmetry groups play a crucial role in their classification."

"Crucial, of course… Hmm, Prof. Priss, I'm afraid it's time to leave now. This is so interesting!"

3:00 P.M. (Jan. 2015, Pisa Headquarters)

Parking near the Agency. "Here we are, Prof. Priss! I'd be delighted to introduce you to my boss. Oh, don't forget your bags."

"Thanks, but I always travel *baggage-free*. And, if you don't mind, I'd like to pay a visit to Galileo's pendulum first. I see the cathedral behind the corner. I'll be back soon."

"You silly!", says my boss, as I complete my report, "you set free the only suspect who confessed his murder! How can you expect him to come back?"

"He will. He has nothing to lose: his worldline is frozen. Like yours and mine, for that matter. We are part of a unique, finitely describable, discrete mathematical structure that includes everything. Free will and the flow of time are pure illusions in the eyes of us frogs—they disappear in the bird's view".

"Birds, frogs, chickens: are you sure Priss is a physicist? Anyway, while you were being indoctrinated by that criminal, I pondered on our phone conversation, and came to some conclusions. Let me tell you.

Your crazy idea was to place a discrete, finite structure, like the input of a computer program, at the bottom of the universe. What's the obvious crazy thing to do next? Look for the right program, run it, and get this universe out! After some googling, I actually found that people like Zuse [11], Fredkin [3], Wolfram [9] have been playing around the idea for decades. For example, programs as simple as elementary cellular automata (CA) can mimic some key features of Nature: not only pseudo-randomness, or deterministic chaos (my favourite!), but also self-similarity and particle interaction."

"I know. For that you need, respectively, automata 30, 18, 110 [9]. But then you get three distinct universes, not one."

"False! Look at Fig. 2. This is three-color CA n. 738926016882 [5]. It proves that you can get those *three* fundamental aspects of physical reality out of *one* deterministic algorithm, even starting with the simplest initial condition! The algorithm is indeed just a function $f : S^3 \rightarrow S$, where $S = \{0, 1, 2\}$ (Appendix 2) so, in a way, Tegmark is right in saying that Nature is math [7], but, I add, *only at the bottom*, and only in an *algorithmic* sense. The algorithm, which need not be a CA [2], starts baggage-free—it initially ignores the fundamental constants of the Standard Model, for example—but *creates* baggage during the journey, by emergence, as the

Fig. 2 Randomness, 'particles' and selfsimilarity in the same CA. Mathematica call: `ArrayPlot[CellularAutomaton[{738926016882, 3, 1}, {{1}, 0}, 880]]`.

computation unfolds. Its function can be written on Priss' T-shirt (Appendix 2), but the computation, with its amazing architecture of emergent phenomena, can't. And I bet it is a non-terminating computation".

"No output?" I ask, thinking of Priss' *total* computable functions.

"No output. Who cares whether the final output is really 42? What matters is what emerges during the trip. By the way, don't forget that, whenever a model of computation is Turing-universal, some of its instances *must* be non-terminating, for some input.[2] Divergence is a virtue, not a sin!"

"Then, if the universe is a finite mathematical structure only at the bottom, how do you explain the unreasonable effectiveness of math in describing *also* the phenomena that emerge during the unfolding?".

"I told you. Math is only *partially* effective in this. Consider again Fig. 2. Math can trivially describe the r.h.s part of the computation, and, with some effort, also the physics of the emergent 'particles' in the middle. But you cannot find a theory for the messy l.h.s component, because that part of the computation is *irreducible*. See, physical theories are nothing but data compressors: they can squeeze huge amounts of experimental data into compact formulas. Not in this case. I mean, the formula is there—function f at the bottom—but there is no way to predict the configuration at step k other than running k steps of the computation. No short-cut.

What we can describe efficiently by mathematics is only the portion of universal algorithmic evolution that is compressible, due to its regularity. Fortunately, it is provably much more likely that something regular and compressible emerge in a computational universe—e.g. in the behaviour of a universal Turing machine running a random program—than from a monkey typing at random on a typewriter [4, 10].

The Universe is not a static mathematical structure—a huge, pre-defined set of elements and relations that we progressively discover. It is the unfolding of a computation,[3] and a relentless source of novelty; its future gifts are mathematically unknowable until they actually come into existence. I have no objection in attributing a respectable status of existence to beautiful pieces of math such as the Mandelbrot set. They did exist at the Plank era, or, if you prefer, they inhabit a platonic, timeless space. I can't say the same about atoms and molecules, frogs and birds, you and me.

In conclusion, no doubt physicists and mathematicians will continue building (or discovering) the world of mathematics, but they should put much more effort in exploring the computational universe too, before we get convinced that chickens are indeed spherical! And for doing this, we need a *gigantic computer*."

"Or a tiny one, as Priss' glasses!" In saying this, I realise that his little jewel is still in my pocket.

---

[2]Diagonal argument: if each instance F_i ($i \in N$) of model F converged on each input $j \in N$, then computable function $G(j) := F_j(j) + 1$ would not be in F, thus F would not be universal.

[3]This conjecture is not to be confused with *the Matrix*—the idea that we live in a simulation.

Appendix 1: Matematica Code for Conway's Tangle Game

```
(* ------ CODE ------ *)

swap[x_] := x + 1
rotate[x_] := -1/x
play[init_, moveList_] := FoldList[#2[#1] &, init, moveList]

(* ------ DATA ------ *)

tangleList = {swap, swap, rotate, swap, swap, swap, rotate,
swap, swap, swap};

untangleList =
{rotate, swap, rotate, swap, swap, rotate,
swap, swap, swap, rotate, swap, swap, swap};

(* ------ RUN ------ *)

play[0, tangleList]
{0, 1, 2, -(1/2), 1/2, 3/2, 5/2, -(2/5), 3/5, 8/5, 13/5}

play[13/5, untangleList]
{13/5, -5/13, 8/13, -13/8, -5/8, 3/8, -8/3, -5/3, -2/3, 1/3,
-3, -2, -1, 0}
```

Appendix 2: The Function for Three-Color CA n. 738926016882

The table below defines function f by providing the value of cell c_i at step t, given the values of cells c_{i-1}, c_i, c_{i+1} at time $t - 1$.

```
{{0, 0, 0} -> 0},
{{0, 0, 1} -> 2},
{{0, 0, 2} -> 2},
{{0, 1, 0} -> 0},
{{0, 1, 1} -> 0},
{{0, 1, 2} -> 1},
{{0, 2, 0} -> 1},
{{0, 2, 1} -> 2},
{{0, 2, 2} -> 2},
{{1, 0, 0} -> 2},
```

```
{{1, 0, 1} -> 2},
{{1, 0, 2} -> 1},
{{1, 1, 0} -> 0},
{{1, 1, 1} -> 0},
{{1, 1, 2} -> 0},
{{1, 2, 0} -> 2},
{{1, 2, 1} -> 2},
{{1, 2, 2} -> 0},
{{2, 0, 0} -> 2},
{{2, 0, 1} -> 2},
{{2, 0, 2} -> 1},
{{2, 1, 0} -> 1},
{{2, 1, 1} -> 2},
{{2, 1, 2} -> 1},
{{2, 2, 0} -> 2},
{{2, 2, 1} -> 0},
{{2, 2, 2} -> 0}.
```

References

1. David Balduzzi and Giulio Tononi. Integrated information in discrete dynamical systems: Motivation and theoretical framework. *PLOS Computational Biology*, 4(6):e1000091, 2008.
2. Tommaso Bolognesi. Algorithmic causal sets for a computational spacetime. In A Computable Universe. H. Zenil (ed.), World Scientific, 2013.
3. Ed Fredkin. Five big questions with pretty simple answers. *IBM J. Res. Dev.*, 48(1):31–45, 2004.
4. Leonid A Levin. Laws of information conservation (nongrowth) and aspects of the foundation of probability theory. *Problemy Peredachi Informatsii*, 10(3):30–35, 1974.
5. Remko Siemerink. *Wolfram Science Summer School 2009 - Favourite three-color cellular automaton*. Pisa, June 2009. http://www.wolframscience.com/summerschool/2009/alumni/siemerink.html.
6. Gerard 't Hooft. The cellular automaton interpretation of quantum mechanics, June 2014. http://arxiv.org/abs/1405.1548 [quant-ph].
7. Max Tegmark. The mathematical universe, 2007. http://arxiv.org/abs/0704.0646 [gr-qc].
8. Max Tegmark. Our Mathematical Universe. Alfred A. Knopf - New York, 2014.
9. Stephen Wolfram. A New Kind of Science. Wolfram Media Inc, 2002.
10. Hector Zenil. The world is either algorithmic or mostly random. *FQXi Essay Contest 2011 - Is Reality Digital or Analog?*, 2011. http://fqxi.org/community/forum/topic/867.
11. Konrad Zuse. Calculating space. Technical report, Proj, MAC, MIT, Cambridge, Mass., 1970. Technical Translation AZT-70-164-GEMIT. Original title: "Rechnender Raum".

The Raven and the Writing Desk

A Dialogue on the Nature of Mathematics and Physical Reality

Ian T. Durham

Abstract In this essay, I use a dialogue between characters from Lewis Carroll's *Alice's Adventures in Wonderland* to discuss the relationship of mathematics to physical reality. In it, I propose that there are two realities: representational and tangible. Mathematics belongs to the former. We can reconcile the two by taking Eddington's stance that the universe is nothing more than our description of it.

The Hatter[1] and the March Hare had just attempted to stuff the Dormouse into the teapot. Finding that he didn't quite fit, they all sat back down, with the exception of Alice who had by that time, tramped off in disgust. The Dormouse immediately resumed dozing whilst the Hatter and the March Hare took to talking once again.

MARCH HARE: [*Absently.*] Why *is* a raven like a writing desk?
HATTER: As I said to that impudent girl, I haven't the slightest idea.
MARCH HARE: A new riddle then!
HATTER: 'Yet what are all such gaieties to me
Whose thoughts are full of indices and surds?
$x^2 + 7x + 53$
$= 11/3$.'[a]
MARCH HARE: That's mathematics. That's not a riddle.
HATTER: According to some fellow named… [*Picks up old book of riddles from table and inspects the cover.*]… Lewis Carroll, it most certainly is a riddle.
MARCH HARE: Lewis Carroll isn't a real person.
HATTER: You mean 'wasn't'. He is deceased.
MARCH HARE: How can he be deceased if he's not a real person?
HATTER: You were the one who said he wasn't real.
MARCH HARE: Well, he's not! …Or wasn't. He was actually Charles Dodgson. Dodgson was real. He was a mathematician, in fact.
HATTER: Ah, well Lewis Carroll was as real as mathematics, then!

[1]In Lewis Carroll's *Alice's Adventure's in Wonderland* he is simply referred to as the 'Hatter' and not the 'Mad Hatter.'.

I.T. Durham (✉)
Department of Physics, Saint Anselm College, Manchester 03102, USA
e-mail: idurham@anselm.edu

A. Aguirre et al. (eds.), *Trick or Truth?*, The Frontiers Collection,
DOI 10.1007/978-3-319-27495-9_6

MARCH HARE: That's absurd. 'Lewis Carroll' is a pseudonym.

HATTER: Mathematics is a pseudonym!

MARCH HARE: For what?

HATTER: Reality!

MARCH HARE: You're mad.

HATTER: So I've been told.

DORMOUSE: [*Sleepily.*] According to the Cheshire Cat, you're *both* mad.

HATTER: Spoken by a fellow who leaves his grin lying round where anyone could make off with it.

DORMOUSE: Perhaps it's a riddle.

MARCH HARE: What's a riddle?

HATTER: [*Crossly.*] Mathematics! Pay attention.

MARCH HARE: Mathematics is a riddle?

HATTER: [*Rolling his eyes.*] The riddle is this: is mathematics real?

MARCH HARE: What do you mean 'is it real?' How could it possibly *not* be real?

HATTER: So your contention is that mathematics is real but a pseudonym is not?

MARCH HARE: You just said that mathematics was a pseudonym for reality. Is a pseudonym the same thing as that which it is pseudonymous of?

DORMOUSE: I can't even parse that sentence.

HATTER: [*To the dormouse.*] Go back to sleep.

MARCH HARE: Reality is what can be measured. I cannot 'measure' Lewis Carroll.

HATTER: I'm a hatter. I measure heads for a living.

MARCH HARE: You're a character in a book!

HATTER: Does that make me any less real?

MARCH HARE: Well, you're less tangible, anyway.

HATTER: Ah, now we're getting somewhere. Does something need to be tangible to be real?

MARCH HARE: [*After some thought.*] It's a different type of reality, I suppose.

HATTER: So we have a tangible reality and a ...

MARCH HARE: Call it... *representational* reality.

HATTER: Is mathematics representational or tangible, then?

MARCH HARE: Representational, I should think. Certainly it is *less* tangible.

HATTER: Less tangible than what exactly?

MARCH HARE: [*Glances round.*] This teapot for instance.

HATTER: That's merely a *description* of a teapot.

MARCH HARE: What do you mean it's merely a description of a teapot?

HATTER: I'm just taking you at your word. We're all characters in a book, are we not?

DORMOUSE: Dialogue, actually.

MARCH HARE: Ah, I think I follow. *This* particular teapot is merely a description of a teapot since it appears in a b–

HATTER: –dialogue.

MARCH HARE: Dialogue. But *some* teapots are actual teapots and not merely descriptions. They are tangible.

HATTER: Precisely.

MARCH HARE: [*Looking satisfied.*] Well, there it is then, isn't it?

HATTER: There *what* is?

MARCH HARE: [*Suddenly looking puzzled.*] What is it we were talking about again?

HATTER: Teapots. Pay attention. Would you allow that this teapot is 'physical?'

MARCH HARE: We just agreed that it's merely a *description* of a teapot. Ergo it cannot possibly be physical.

HATTER: [*Takes out reading glasses and a dusty, green book.*] Ahem. 'Physical knowledge… has the form of a description of a world. We *define* the physical universe to be the world so described.'[b]

MARCH HARE: [*Narrowing his eyes.*] What are you playing at?

HATTER: Merely this: the physical universe is that which is described by physical knowledge.

MARCH HARE: That *universe* which is therein described?

HATTER: That *anything* which is therein described.

MARCH HARE: So the universe is simply a description?

HATTER: The universe *is* the sum of our knowledge *of* it. That knowledge takes the form of a description that is ultimately of a mathematical nature.

MARCH HARE: Wait a moment. Yes, wait right there. Are you actually saying that the universe is nothing but mathematics?

HATTER: What I am saying is that the *description* of the universe is mathematical.

MARCH HARE: But you just said that the universe *is* our description *of* it! Oh, I'm so confused!

DORMOUSE: *I'm* sleepy.

HATTER: What I said was this: the universe *is* the sum of our knowledge *of* it. The form that knowledge takes is mathematical. There is a difference.

MARCH HARE: No, still confused.

HATTER: You agree that there are two realities: representational and tangible?

MARCH HARE: Reluctantly.

HATTER: Our knowledge of the universe, in the form of mathematics, is representational. The universe itself is tangible.

MARCH HARE: But, shouldn't each type of reality be equivalent? For instance, there are plenty of things I can *represent* that cannot possibly be *tangible*.

HATTER: [*Aside.*] Like talking rabbits?

MARCH HARE: Yet everything that is tangible may be represented. Take infinity, for instance. Infinity is a representation, but it cannot possibly represent anything tangible.

HATTER: How so?

MARCH HARE: Well, you can't count to infinity. It is a Sysyphean task. There is nothing tangible about infinity.

HATTER: Neither is there anything particularly tangible about 'two' or 'ten.' They are abstract concepts. You cannot measure the concept 'two.'

MARCH HARE: Ah, but I *can* measure two *of* something. I cannot measure an infinite amount of *anything*!

HATTER: But how do you *know* that?

MARCH HARE: [*Turning red.*] Because I haven't the time!

HATTER: This is a b-

MARCH HARE: Dialogue.

HATTER: Dialogue. Remember that it is always six o'clock.

MARCH HARE: Well, ever since the Queen accused you of murdering Time.

HATTER: Time is meaningless.

MARCH HARE: But only because the Queen thinks you murdered him!

HATTER: Indeed. He no longer responds to any of my letters! Ergo, he must be meaningless.[c]

MARCH HARE: Page numbers aren't.

HATTER: An experiment then!

MARCH HARE: I am wholly in favor of experiments!

HATTER: [*Looks at the Dormouse.*] Start counting.

DORMOUSE: [*Yawning.*] One…

[a][1].
[b][2].
[c]Time is a 'person' in *Alice's Adventures in Wonderland.*

At that moment, the White Rabbit hurries by, pulling a watch from his waistcoat pocket, all the while mumbling, 'Oh my ears and whiskers, how late it's getting!'

MARCH HARE: That fellow is always in a hurry.

HATTER: [*Ignoring the March Hare's comment about the White Rabbit entirely.*] Fine. Suppose I grant you that it is impossible to measure an infinite amount of anything. You would agree, I assume, that infinity is nevertheless representational?

MARCH HARE: It is self-evident.

DORMOUSE: Two…

HATTER: 'Two' is representational.

MARCH HARE: Also self-evident.

HATTER: There is clearly a difference between 'two' and 'infinity.' As you said, you can count to two but you can't count to infinity.

MARCH HARE: [*Looking suddenly defeated.*] And here I thought we were making progress.

DORMOUSE: Three…

HATTER: Not nearly as much as you might think. Counting (or rather, attempting to count) to infinity is an example of a *potential* infinity, not an *actual* infinity.

MARCH HARE: Oh, my head hurts!

HATTER: Actual infinity assumes that mathematical objects can form an actual, completed totality of something.

MARCH HARE: This conversation is absurd.

HATTER: *All* our conversations are absurd. Now pay attention. Counting numbers is an example of a non-terminating process that produces an unending, infinite sequence of results. Yet each individual result is finite and is reached in a finite number of steps. Hence it is a *potential* infinity.

DORMOUSE: Four…

HATTER: You see? He's only at four, but if we leave him here long enough he has the *potential* to reach infinity.

MARCH HARE: [*Timidly.*] Because… he's merely representational?

HATTER: Oh, I have no idea. He could be tangible for all I know. The fact is that he is attempting to count to infinity. Barring any interference from anything else–

MARCH HARE: Like the end of this dialogue.

HATTER: –or the end of the universe—he will reach his goal.

MARCH HARE: Well, I suppose...

HATTER: Hence, we refer to that as a *potential* infinity. Now consider integers.

MARCH HARE: What about them?

HATTER: We use the word 'integers' (in the plural) to refer to integers in general, not just as an abstract concept (in which case we would likely use the singular 'integer'), but as the totality of all integers.

MARCH HARE: I think I see where this is headed.

HATTER: But there are an infinite number of integers. So when we speak of 'integers' in the generic, plural sense, we are referring to a totality of all integers, of which there are infinitely many. That is an example of an *actual* infinity.

MARCH HARE: No… no, you've lost me.

HATTER: [*Gives the March Hare a rather cross look.*] When I say the word 'integers'—plural, mind you!—what is the first thing that comes to mind?

DORMOUSE: Forty-two…

MARCH HARE: How did he reach forty-two so quickly? It seems as if he was just at four a moment ago.

HATTER: This is a b–

MARCH HARE: Dialogue.

HATTER: *–dialogue*. We are stuck at six o'clock, remember? Time has no meaning.

MARCH HARE: Ever since the Queen—

HATTER: Focus! Now, what is the first thing that comes to mind when I say the word 'integers?'

MARCH HARE: Numbers. Lots of them.

HATTER: *All* of them, I should think. Or all the 'counting' numbers anyway.

MARCH HARE: And their negatives, of course.

HATTER: If you wish to be pedantic about it.

MARCH HARE: I insist. You said 'integers' not 'natural numbers.'

HATTER: Fine.

MARCH HARE: So when you say 'integers' I will grant that you are referring to all such numbers.

HATTER: Most of the time.

MARCH HARE: Most… wait, what?

HATTER: I am merely allowing for vagaries in the English language. Perhaps these problems do not exist in German.

DORMOUSE: There's a word for *everything* in German!

HATTER: Not quite. For the time being German is still a finite language. Keep counting. [*To the March Hare.*] Now, how many integers are there?

MARCH HARE: Well, an infinite number, of course!

HATTER: Exactly. Thus, the collection of all integers is an example of *actual* infinity.

MARCH HARE: Ah, I see… I think.

HATTER: Now what is the first thing that comes to mind when I say 'real numbers?'

MARCH HARE: This is like a linguistic Rorschach test. What if I simply say 'potato?'

HATTER: I will send you off to the Queen of Hearts. [*The March Hare instinctively reaches for his neck.*] Now *focus*! What comes to mind when I say 'real numbers?'

MARCH HARE: Well… numbers, of course!

HATTER: *And*…?

MARCH HARE: Oh, I see what you're playing at. There are an infinite number of real numbers. It's another example of an actual infinity.

HATTER: So there are just as many integers as there are real numbers?

MARCH HARE: Absolutely. Infinity is infinity.

HATTER: How many integers are there between zero and one-hundred, inclusive?

DORMOUSE: One-hundred…

MARCH HARE: [*Nods in the direction of the Dormouse.*]

HATTER: And how many real numbers are there between zero and one-hundred, inclusive?

MARCH HARE: [*Opens his mouth to speak, then shuts it with a puzzled look on his face, then opens it again to speak.*] Wha…hang on.

HATTER: *How many*!

MARCH HARE: Well, you don't have to be so rude about it. I suppose I might admit that there are an infinite number of real numbers between zero and one-hundred.

HATTER: So are there as many integers as there are real numbers?

MARCH HARE: [*Looking puzzled.*] Well… no, I suppose not.

HATTER: Precisely! There are multiple infinities, all of different sizes.

MARCH HARE: But, that's ridiculous!

HATTER: No more ridiculous than characters from a book (one of which is a rabbit, I might add) having a conversation about the nature of mathematics.

MARCH HARE: I'm a hare. That fellow with the pocket watch who just ran by is a rabbit. And characters from books have these sorts of conversations all the time. It is still ridiculous.

HATTER: Ah, so you're an intuitionist, then?

MARCH HARE: A what?

HATTER: Intuitionist. You accept potential infinities but not actual infinities.

MARCH HARE: I suppose.

HATTER: The Queen of Hearts has a scroll of names. Are you familiar with it?[a]

MARCH HARE: Oh yes! One does not want to find one's name on that scroll!

HATTER: How many names are on that scroll?

DORMOUSE: One-hundred thirty seven...

MARCH HARE: I can't say I've ever seen the beginning or the end of that scroll. It seems to go on forever.

HATTER: So it might be infinite?

MARCH HARE: Absurd!

HATTER: Well, look, we know my name is on that scroll somewhere.

MARCH HARE: Ever since the Queen accused you of murdering–

HATTER: Enough already!

MARCH HARE: Suffice it to say your name is on the scroll.

HATTER: Yes, well the White Rabbit is the Court Herald, so suppose that we put him on the task of finding my name on that scroll. How would he do so?

MARCH HARE: He would begin by reading the names on the scroll.

HATTER: Where would he begin?

MARCH HARE: Anywhere!

HATTER: And so, not knowing where in the execution order I am, he could read names in either direction, correct?

MARCH HARE: Correct.

HATTER: But it takes a finite amount of time to read each name.

MARCH HARE: I thought it was always six o'clock?

HATTER: That's only for us. Pay *attention*! Now, it takes a finite amount of time for each name to be read. Therefore, we shall never *know* that the scroll is infinite in either direction!

MARCH HARE: Exactly! I *knew* you'd come round!

HATTER: Come round to what? I am merely confirming that you are an intuitionist. You only believe in potential infinities.

MARCH HARE: No need to get my hopes up then.

DORMOUSE: One-thousand nine-hundred seventy-four...

MARCH HARE: Good heavens, he's sped up quite a bit! I think he's skipping numbers.

HATTER: Even if he skipped every third number—or even if he skipped all numbers *except* every third number—there are still an infinite number of numbers for him to count.

MARCH HARE: But how does all this relate to the reality of mathematics?

HATTER: *Pay attention*!! We have agreed that there are multiple sizes of infinity, correct?

MARCH HARE: Integers, real numbers, and all that?

HATTER: [*Nods.*] And we have agreed that there are two *types* of infinity, correct?

MARCH HARE: Potential and actual.

HATTER: The Dormouse here is counting the positive integers.

MARCH HARE: Natural numbers. At least some of them.

HATTER: Naturally. We have already established that the set of all positive integers is an actual infinity.

MARCH HARE: I suppose…

HATTER: But, as the Dormouse is attempting to count them, they represent a potential infinity.

MARCH HARE: Absolutely.

HATTER: Well, which is it? Or… is it both?

MARCH HARE: I see what you're doing. You're trying to get me to admit that there is only one type of infinity.

HATTER: Merely that they are equivalent.

MARCH HARE: I still don't see what this has to do with the reality of mathematics.

HATTER: [*Turning red.*] *They're all just descriptions! PAY ATTENTION!!!* Everything is merely a description! Potential, actual—we're simply describing things!

MARCH HARE: Some of which are tangible and some of which are representational.

HATTER: [*Bursting with excitement.*] YES! The universe *is* the sum of our knowledge *of* it! Nothing more, nothing less!

MARCH HARE: And mathematics is representational?

HATTER: Precisely.

MARCH HARE: But if the universe is merely description, then it doesn't matter.

HATTER: After all, just because I'm a character in a *dialogue* doesn't make me any less a part of the universe.

[a]This is entirely my own invention and does not appear in *Alice's Adventures in Wonderland*. However, the idea I outline here is based on Kleene's description of a Turing machine tape [3].

There was a long pause, during which the Dormouse's counting slowly turned into snoring.

MARCH HARE: No. I simply don't buy it. I'll stick with tangible reality, thank you very much.

HATTER: A character in a dialogue has chosen tangible reality. How comforting.

MARCH HARE: Perhaps if I were in a book I would have chosen differently.

HATTER: Perhaps if *I* were in a book I would have sent you to the executioner.

MARCH HARE: So that's it then. We have a conclusion, have we?

HATTER: You're all about conclusions.

MARCH HARE: If I were *all* about conclusions, I would demand an answer to that riddle about the raven and the writing desk.

HATTER: Oh, there's an answer to that.

MARCH HARE: But you said…

HATTER: So I did, didn't I?

MARCH HARE: Never mind. Some things are best left unknown.
HATTER: Like where my name is on that scroll.
MARCH HARE: Still, I'd like to conclude this tea party even if it means finding your name on the scroll.
HATTER: Let's say half-past six?
MARCH HARE: Marvelous plan.
HATTER: Let's all move one place on.[a]
DORMOUSE: [*A snore slowly turns into a word...*] Infinity. [*Suddenly realizes what he's done and his eyes nearly pop out of his tiny head.*] I've...I've done it!
HATTER: Done what?
DORMOUSE: I've counted to infinity!
MARCH HARE: There's a lesson in that somewhere.
HATTER: No there isn't. *TEAPOT*!!

[a]See Chapter VII: A Mad Tea-Party in *Alice's Adventures in Wonderland*.

The Hatter and the March Hare then grab the Dormouse and attempt to stuff him into the teapot.

References

1. Lewis Carroll. *Rhyme? and Reason?*, Chapter: Four Riddles, I. MacMillan and Co., London, 1883.
2. Arthur Stanley Eddington. *The Philosophy of Physical Science*. Cambridge University Press, Cambridge, 1939.
3. Stephen C. Kleene. *Introduction to Metamathematics*. North-Holland Publishing Company, Amsterdam, NY, 1971.

The Deeper Roles of Mathematics in Physical Laws

Kevin H. Knuth

Familiarity breeds the illusion of understanding

–Anonymous

Abstract Many have wondered how mathematics, which appears to be the result of both human creativity and human discovery, can possibly exhibit the degree of success and seemingly-universal applicability to quantifying the physical world as exemplified by the laws of physics. In this essay, I claim that much of the utility of mathematics arises from our choice of description of the physical world coupled with our desire to quantify it. This will be demonstrated in a practical sense by considering one of the most fundamental concepts of mathematics: additivity. This example will be used to show how many physical laws can be derived as constraint equations enforcing relevant symmetries in a sense that is far more fundamental than commonly appreciated.

Introduction

Many have wondered how mathematics, which appears to be the result of both human creativity and human discovery, can possibly exhibit the degree of success and seemingly-universal applicability to quantifying the physical world as exemplified by the laws of physics [17, 20]. That is, if the laws of physics are taken as fundamental, then how can it be that mathematics, which is often developed as a creative act, can possibly be relevant to the physical world? Or is mathematics somehow more fundamental than physics, and are we instead discovering the laws

K.H. Knuth (✉)
Departments of Physics and Informatics, University at Albany (SUNY), Albany, NY, USA
e-mail: kknuth@albany.edu

A. Aguirre et al. (eds.), *Trick or Truth?*, The Frontiers Collection,
DOI 10.1007/978-3-319-27495-9_7

of mathematics and identifying which ones apply to or perhaps underlie particular physical situations?

These questions reside alongside longstanding questions regarding the nature of physical law. For example, Isaac Newton introduced the idea that some physical laws can be derived from other more fundamental physical laws—often given the more distinct title of *principles*. Today, many of these principles take the form of conservation laws or symmetries, which helps to account for their universality. However, this raises many questions, such as whether there exists a unique minimal set of fundamental principles from which everything derives, or whether some physical laws are derivable and others are determined by chance or decree.

Despite this, we can already see that mathematics plays at least two roles. The first role is related to symmetries, which necessarily represent foundational concepts since they are not readily derivable from more fundamental concepts. The second role is related to calculation where equations are used to quantify physical phenomena. While some of the equations are known to be derivable from more fundamental principles, some have been adopted as foundational concepts in their own right, such as specific Lagrangians. This provides an important clue, especially since we know from experience that many equations have been arrived at through educated guesses or in some cases even trial-and-error.

In this essay, I will argue that much of the mathematics that applies to physics arises from our choice of description of the physical world coupled with our desire to quantify it. I will demonstrate this in a practical example by considering one of the most fundamental concepts of mathematics: additivity. This explicit demonstration reveals that there are two distinct aspects to the role that mathematics plays. The first aspect is related to ordering and associated symmetries, and the second aspect is related to quantification and the equations that enable one to quantify things. The result is that to a great degree the mathematical equations that we consider to represent physical laws arise as constraint equations that enforce basic symmetries resulting from our chosen descriptions and desire to quantify. While it may be that not all equations can be arrived at in this way, by showing that this is more widespread than commonly appreciated, I aim to suggest that this is the reason why mathematics is so effective.

Questions

When I was a physics student, I was once troubled by a seemingly simple question. I went around the department posing my question to fellow students as well as professors. Most people seemed to feel that what I was asking was a silly question and some had quick answers. However, despite these situations, I never received what I felt to be a satisfactory or insightful response.[1] I had asked:

[1]In preparing for this essay, I was pleased to find that Hamming had posed a similar question: "I have tried, with little success, to get some of my friends to understand my amazement that

> Why is it that when I take two pencils and add one pencil, I always get three pencils? And when I take two pennies and add one penny, I always get three pennies, and so on with rocks and sticks and candy and monkeys and planets and stars. Is this true by definition as in 2+1 defines 3? Or is it an experimental fact so that at some point in the distant past this observation needed to be verified again and again?

The fact that this simple question is focused on a very familiar problem obscures the fact that it embodies the very same wonderment that many of us possess when considering the "unreasonable effectiveness" of mathematics in physics. I have always felt that if we can come to understand the answer to this simple question, then we ought to begin to be able to understand the role that mathematics plays in physics at a whole.[2] On the other hand, if this question leads us to running in circles then we clearly have a long way to go.

Now some may scoff at this question, as many of my fellow students and professors did. One common answer was that this is just measure theory: "You have sets where the cardinality of the set is a measure, and measures on sets sum when you combine them via set union." However, what many respondents failed to realize was that this is an *axiom* of measure theory, which means that it is *assumed*. As a physicist interested in the foundations, assumptions cause me concern. And despite its familiarity, this is a big assumption indeed! Furthermore, taking this to be the answer to my query simply transforms my question to: "Why must set measures be additive?"

The answer to this is not obvious,[3] and it is central to the problem of the unreasonable effectiveness of mathematics. Clearly addition works, and this might suggest that the mathematics we use has been selected because it has worked every time we have employed it, which would make the use of addition an experimentally-observed result. Now, it could very well be that this is how this all came about historically. That is, it could be that the mathematics that humanity developed was selected (evolved) to work again and again without violation.

But why all the fuss; the original question is just about counting isn't it?

Could it really be the case that this goes deeper?

(Footnote 1 continued)
the abstraction of integers for counting is both possible and useful. Is it not remarkable that 6 sheep plus 7 sheep make 13 sheep; that 6 stones plus 7 stones make 13 stones? Is it not a miracle that the universe is so constructed that such a simple abstraction as a number is possible? To me this is one of the strongest examples of the unreasonable effectiveness of mathematics. Indeed, I find it both strange and unexplainable" [17].

[2]Einstein expressed a similar sentiment about particle physics: "I would just like to know what an electron is."

[3]This is especially difficult since additivity of set measure is assumed by mathematicians. That is where they decided to start. Where else would one start? This highlights one of the subtle and insidious difficulties faced by those in foundational studies. Once it is decided to adopt an assumption as a foundational construct it precludes the ability to delve deeper within that framework. Furthermore, it discourages others from doing so through what I call "The Curse of Familiarity" where one's familiarity with a problem fosters an illusion of understanding that blinds one from seeing subtle clues, hints, connections and/or difficulties. The parable of Newton and the apple is an example where Newton momentarily saw through the familiarity of falling apples to realize a connection with the falling of the Moon about the Earth.

Symmetries and Order

There are some important symmetries at play when I combine sets of pencils or pennies or monkeys or stars.[4] If I combine set A with set B to form the joint set D, the result will be the same as if I combine set B with set A. This symmetry is called *commutativity*. Considering sets joined by set union, since $D = A \cup B$ and $D = B \cup A$, we can write

$$A \cup B = B \cup A. \qquad \textbf{Commutativity}$$

Now there is another important symmetry called *associativity* where the order in which one unites sets of objects also does not affect the final results. That is, in combining three sets A, B, and C, I could first combine A with B and then combine the result with C, or I could combine A with the result of the combination of B with C, and so on. In mathematical parlance, we have that

$$(A \cup B) \cup C = A \cup (B \cup C) = (A \cup C) \cup B, \qquad \textbf{Associativity}$$

so that we can just drop the parentheses and simply write $A \cup B \cup C$.

I feel relatively confident that these conceptual symmetries are, and need to be, experimentally observed when combining many things such as pencils and monkeys and so on. This is mainly because there are some examples where these symmetries are not experimentally observed. For example, if I combine a burning match with a paper napkin and then sometime later combine them with a glass of water, I get a very different result than if I were to first combine the napkin with the glass of water and then combine those with the burning match. Of course, this depends upon our chosen level of description since if we included the air and counted the atoms, we would expect the results to be the same. This relies on the concept of closure where combining things results in the same sorts of things.

Sets are also interesting because they can be ordered. Sets can include other sets, and in fact we see this in the simple example above where the set $A \cup B \cup C$ includes the set $A \cup B$. In mathematical parlance, we say that the set $A \cup B$ is a subset of the set $A \cup B \cup C$ and write

$$A \cup B \subseteq A \cup B \cup C \qquad \textbf{Subset Inclusion}$$

This leads to a hierarchy[5] of sets, which can be illustrated using what are called Hasse diagrams in order theory. Figure 1a illustrates this relationship among sets. The set union of two sets can be found by starting at each set and moving upward in the diagram until the two paths converge. Similarly, the set intersection can be found by starting at two sets and moving downward until the paths converge.

[4]Finite sets are sufficient for our purposes here as we are not attempting to model an infinite world.

[5]It would be better to call it a heterarchy since sets in general cannot be linearly ordered.

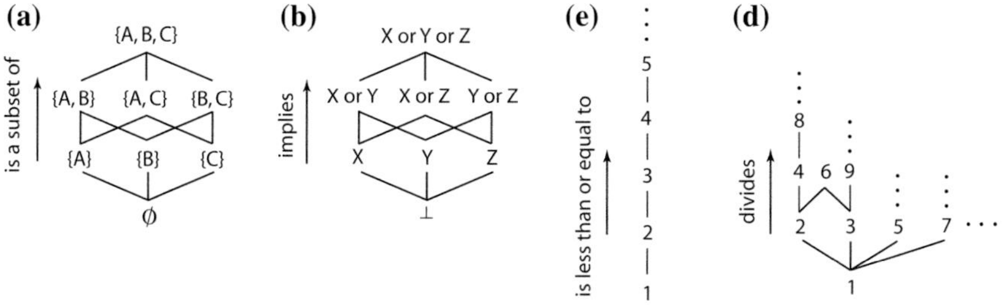

Fig. 1 **a** The lattice of sets A, B, C ordered by set inclusion. The lattice meet and join are the operations set intersection and union, respectively. The bottom element is the null set. This structure is called a Boolean lattice as it represents a Boolean algebra. **b** The lattice of logical statements X, Y, Z ordered by implication also forms a Boolean lattice with the meet and the join being the logical 'and' and 'or' operations, respectively. The bottom element is the falsity. **c** The counting numbers ordered by the usual is-less-than-or-equal-to forms a lattice with the meet and join being the min and max functions. **d** However, the counting numbers can also be ordered according to whether one divides another with the meet and join being the GCD and LCM

Order theory focuses on the ways sets of objects, called elements, can be ordered [2, 5]. The useful concept is that of a partially ordered set of elements. A *partially ordered set* (or *poset* for short) is defined as a set of elements that can be compared with a binary ordering relation, generically denoted $\leq$, that exhibits the properties of reflexivity, antisymmetry and transitivity (see Technical Endnotes).

The sets A, B, C above along with the binary ordering relation of subset inclusion $\subseteq$ form a poset. However, this poset has additional properties that give it the special name of a lattice. Specifically, each pair of elements x and y has a unique least upper bound, $x \vee y$, called the join, and a unique greatest lower bound, $x \wedge y$, called the meet, such that the join and the meet are associative. So that in the case of our sets ordered by subset inclusion, we have that the lattice join is the set union, and the lattice meet is the set intersection.

There are many other posets which form lattices. Figure 1 illustrates a few familiar examples. As described earlier, the first example in Fig. 1a is the lattice of sets ordered by set inclusion. The second example (Fig. 1b) shows the set of logical statements X, Y, Z ordered by logical implication where the lattice join is the logical "or" and the lattice meet is the logical "and". Figure 1c, illustrates the counting numbers ordered by the usual less-than-or-equal-to $\leq$. This forms what is called a totally ordered set or a chain, where the lattice join and meet are the *max* and *min* functions, respectively. For example, $1 \vee 2 \equiv \max(1, 2) = 2$ and $3 \wedge 4 \equiv \min(3, 4) = 3$. Finally, Fig. 1d illustrates that the counting numbers can be ordered in another way leading to a different poset. Here they are ordered according to whether one number divides another. The lattice join is given by the Least Common Multiple (*LCM*) and the lattice meet is the Greatest Common Divisor (*GCD*).

Now, what is important in all of this is that the simple act of ordering objects and the properties of commutativity and associativity allows one to view things in two different ways. On one hand we have a sort of hierarchy of elements determined by the binary ordering relation. And on the other hand, we can consider the join and

(a) Sets, $\subseteq$ $A \subseteq B \;\Leftrightarrow\; \begin{matrix} A \cap B = A \\ A \cup B = B \end{matrix}$	**(c)** Counting Numbers, $\leq$ $p \leq q \;\Leftrightarrow\; \begin{matrix} \min(p,q) = p \\ \max(p,q) = q \end{matrix}$
(b) Logical Statements, implies $X \textit{ implies } Y \;\Leftrightarrow\; \begin{matrix} X \textit{ and } Y = X \\ X \textit{ or } Y = Y \end{matrix}$	**(d)** Counting Numbers, divides $p \textit{ divides } q \;\Leftrightarrow\; \begin{matrix} \mathrm{GCD}(p,q) = p \\ \mathrm{LCM}(p,q) = q \end{matrix}$

Fig. 2 This figure illustrates how the ordering relation relates to the induced algebra for the four lattices illustrated in Fig. 1. Later we will see that the fact that each of these algebras (set union and intersection, logical "and" and "or", min and max, and GCD and LCM) is associative, means that the same law (Sum Rule) will emerge when we consistently quantify the elements

meet to be algebraic functions that take two elements to a third. The result is that a lattice is an algebra, and the ordering relation is related to the algebraic relation by what is called the consistency relation

$$a \leq b \quad \Leftrightarrow \quad \begin{matrix} a \wedge b = a \\ a \vee b = b \end{matrix} \qquad \textbf{Consistency Relation}$$

With these concepts in mind, connections can now be made. Figure 2 shows the consistency relation written specifically for each of the four posets illustrated in Fig. 1. One can now see that ordering, commutativity and associativity underlie a class of universal phenomena. I will next discuss how this leads to mathematics which gives rise to physical laws with a degree of universal applicability.

Quantification

Let us now consider how we might go about quantifying elements of lattices. Two of the four examples above involve numbers directly and thus appear to be easy. There is nothing wrong with easy. Such cases will help us find the way since if there exists a general rule, it must apply to special cases.

We begin with the concept of quantification. The idea here is very simple. To each element p we will assign a numeric value $v(p)$. Now, if we want our quantification scheme to maintain some representation of the ordering relation then for elements p and q where $p \geq q$ we assign values such that $v(p) \geq v(q)$. Essentially, here we are mapping elements to a total order—thus ranking them. This puts a strong constraint on the values we can assign.

Now given two disjoint elements x and y (which have null meet) and the values assigned to them $v(x)$ and $v(y)$, it would be good to know what number $v(x \vee y)$ we should assign to their join $x \vee y$. Clearly $v(x \vee y)$ must be greater than or equal to both $v(x)$ and $v(y)$. But can it be *any* such number? How can we be sure to avoid conflicts and ensure consistency with other joins in the same lattice?

First, if we want the assigned quantification to encode the underlying relationship, then we must assume that the number $v(x \vee y)$ we assign to the join $x \vee y$ of two disjoint elements, x and y, is a function of the numbers $v(x)$ and $v(y)$. That is, we should be able to write

$$v(x \vee y) = v(x) \oplus v(y)$$

where $\oplus$ is a real-valued binary function to be determined. What could $\oplus$ possibly be? What would work and what would not work? For example, it couldn't be the min function since $\min(v(x), v(y))$ would result in picking the smallest number, which would violate the fact that the quantification must be greater than or equal to the largest number. The max function will not do either, since we will end up with a degenerate measure that simply ignores the lesser component, which is contrary to the goal of ranking.[6] The lesson is that not just any function will do.

Consider the symmetries of commutativity and associativity. First, since commutativity says that $x \vee y = y \vee x$ we have simply that their quantifications are equal $v(x \vee y) = v(y \vee x)$ which means that the operator $\oplus$ must be commutative: $v(x) \oplus v(y) = v(y) \oplus v(x)$. Writing $a = v(x)$ and $b = v(y)$, the commutative structure is more clear:

$$a \oplus b = b \oplus a. \qquad \textbf{Commutativity of } \oplus$$

Furthermore, since the lattice join is associative, the operator $\oplus$ must also be associative:

$$(a \oplus b) \oplus c = a \oplus (b \oplus c), \qquad \textbf{Associativity of } \oplus$$

where I have introduced a third element z disjoint from x and y, which has an associated value $c = v(z)$. The equation above is a functional equation called the Associativity Equation where the aim is to determine the function $\oplus$. The solution to this functional equation is known to be such that the function $\oplus$ must be an invertible transform of addition [1, 4, 10]. That is $\oplus$ must take the form:

$$a \oplus b = f^{-1}(f(a) + f(b))$$

where the function f is an arbitrary invertible function. In terms of the values we assigned to the elements this can be re-written as

$$f(v(x) \oplus v(y)) = f(v(x)) + f(v(y)),$$

---

[6]Since the goal is to rank elements via quantification (mapping elements to a total order), it may be helpful to formalize the desired systematic preservation of inequality by making explicit the assumption of cancellativity where for disjoint elements x, y, and z, where $v(x) \leq v(y) \leq v(z)$, we have $v(x) \oplus v(z) \leq v(y) \oplus v(z)$, which is implicit in any generally-useful (non-degenerate) notion of ranking.

which is

$$f(v(x \vee y)) = f(v(x)) + f(v(y)).$$

This is significant because we can simply perform a *regraduation* on the valuations v by instead assigning different values $u(x)$ defined by $u(x) = f(v(x))$ so that the quantification is additive for disjoint x and y

$$u(x \vee y) = u(x) + u(y). \qquad \textbf{Additivity}$$

This means that any consistent quantification of the elements of a lattice can be transformed into one that is additive. The fact that we can *always* do this means that order (cancellativity), commutativity and associativity results in additive measures. We have, in fact, *derived* the countable additivity axiom of measure theory from a deeper symmetry principle!

The Ubiquity of Additivity

We now can understand why we add quantities when we combine things. It doesn't always work. The properties of order, commutativity and associativity must apply to the selected description. So now when I take a pair of pencils and combine them with another pencil, I can quantify the union of these sets of pencils by simple addition. This works because sets of pencils are closed and can be ordered, and combining pencils is commutative and associative. This is the answer to my simple question in graduate school (and Hamming's question). And now we will see that the consequences are quite profound.

The additive rule above was derived for disjoint elements. One can show that in general (see Technical Endnotes) the additive rule is

$$u(x \vee y) = u(x) + u(y) - u(x \wedge y),$$

where we have to subtract off the value assigned to the meet of the two elements to avoid double-counting [8, 9]. In order theory, this is known as the *inclusion-exclusion principle* [7, 8]. Others simply call it the *Sum Rule*. What is remarkable is that the Sum Rule appears over and over again, and now we can understand why. This is a consequence of closure, ordering, commutativity and associativity.

Table 1 lists several examples of the Sum Rule in a variety of applications. The first four examples correspond to the four lattices in Fig. 1. The example dealing with measures on sets is quite broad and includes physical volumes, surface areas, mean lengths, as well as linear superposition of potentials and other scalar quantities. The second example arises from the fact that probability quantifies the degree to which one logical statement implies another [10]. This has enormous implications for physics since probability is one of the foundational concepts underlying both statistical mechanics and quantum mechanics. The third example, known as Polya's

Table 1 An illustration of the ubiquity of the sum rule

$m(A \cup B) = m(A) + m(B) - m(A \cap B)$	Measures on sets
$P(A\, or\, B\|I) =$ $P(A\|I) + P(B\|I) - P(A\, and\, B\|I)$	Probability theory
$\max(a, b) = a + b - \min(a, b)$	Polya's Min-Max rule
$\log(LCM(p, q)) =$ $\log p + \log p - \log(GCD(p, q))$	Integral divisors
$I(A; B) = H(A) + H(B) - H(A, B)$	Mutual information
$\chi = V - E + F$	Euler characteristic
$E = (A + B + C) - \pi$	Spherical excess
$I_3(A, B, C) = \|A \sqcup B \sqcup C\| - \|A \sqcup B\| -$ $\|A \sqcup C\| - \|B \sqcup C\| + \|A\| + \|B\| + \|C\|$	Three slit problem

Min-Max Rule [18], is one of my favorites due to its simplicity and elegance. To find the larger of two numbers, sum them and take away the smallest! The fourth example illustrates that products are essentially linear (in the log) and they also arise from these basic symmetries [10].

Next, the definition of Mutual Information is a sum and difference of entropies, and is central to Information Theory [3]. This is followed by two less obvious examples related to geometry [7]. The Euler Characteristic of a regular polytope is found by taking the number of Vertices minus the number of Edges plus the number of Faces. The spherical excess is an important quantity in spherical geometry where A, B, and C are the angles of a spherical triangle.

The last example is the formula required to compute the quantum amplitude of the three slit problem where the square-cup operators represent combinations of slits [19]. Here the quantity being summed is a complex number. Regardless of this, the formula still arises from associativity. It may be surprising, but one can *derive* the Feynman Rules for combining quantum amplitudes by relying on symmetries, such as associativity and distributivity along with consistency with Probability Theory [15, 16]. And more recently, many of these same symmetries were used to derive the mathematics of flat spacetime as constraint equations describing the consistent quantification of a partially ordered set of events [11–13] as well as to derive the quantum symmetrization postulate [14].

Now that we see how the quantification of phenomena is constrained by symmetries, and that this is far more widespread than the usual conservation laws and symmetries considered in physics, we can better understand precisely how mathematics is related to physical laws. The following questions now arise: "How far does this go?" and "To what degree are physical laws derivable and to what degree are they accidental, contingent or decreed by Mother Nature?"

Conclusions

The results here shed light on the long-standing questions surrounding the unreasonable effectiveness of mathematics. We have seen interplay between two different aspects of mathematics. The first aspect is related to ordering and symmetries, and the second aspect is related to quantification and the equations that enable one to quantify things. Our choices in the phenomena that we focus on, the descriptions we adopt and the comparisons that we find important often amount to selecting a particular concept of ordering,[7] which can possesses symmetries. The ordering relation and its symmetries in turn constrain consistent attempts at quantification resulting in constraint equations, which in many cases are related to what are considered to be physical laws. Much of the wonderment surrounding the unreasonable effectiveness of mathematics is not associated with the first aspect of ordering and symmetries since these more clearly depend on a choice of description and comparison, which in turn results in symmetries that can be easily observed and verified. Instead such wonderment is associated with the fact that we have equations that consistently allow us to quantify the physical world, and that these equations not only work very well, but in many cases exhibit some degree of universality. If we consider the equations themselves to be fundamental then the success of mathematics is somewhat of a mystery. But if we step back and release ourselves from familiarity and consider order and symmetry to be fundamental, then we see these equations as rules to constrain our artificial quantifications in accordance with the underlying order and symmetries of our chosen descriptions.

In some sense, this should not be surprising since it has been generally believed that the laws of physics reflect an underlying order in the universe. In fact, here it is explicitly demonstrated that some laws of physics not only reflect such order, but in fact can be derived directly from it. This has enormous implications for the direction and progress of foundational physics in the sense that it enables one to see that common mathematical assumptions, such as additivity, linearity, Hilbert spaces, etc., while familiar, are most likely not fundamental. Instead, there is room to delve deeper by identifying the fundamental symmetries and order-theoretic concepts that underlie physical theories.

The universality of mathematics, and specific mathematical relations, has been thought by many to be mysterious. Hamming wrote the following of Wigner's exposition [20]:

> Wigner also observes that *the same mathematical concepts* turn up in entirely unexpected connections. For example, the trigonometric functions which occur in Ptolemy's astronomy turn out to be the functions which are invariant with respect to translation (time invariance). They are also the appropriate functions for linear systems. The enormous usefulness of the

[7]Do we select a particular concept of ordering? Of course we do; an example is given in Fig. 2 where the counting numbers are ordered in two different ways. Selecting one way of ordering gives you one set of laws (min and max) and selecting the other gives you another set of laws (GCD and LCM). The entire of field of number theory results as an attempt to study relationships between these two resulting sets of laws.

> same pieces of mathematics in widely different situations has no rational explanation (as yet) [17].

As demonstrated in this essay, I believe that the answer lies in the deeper symmetries that various problems exhibit. To quote Jaynes:

> the essential content ... does not lie in the equations; it lies in the ideas that lead to those equations [6].

In addition to mystery and wonderment, this universality of mathematics has also caused a great deal of confusion. For example, some believe that the theory of quantum mechanics represents some kind of generalized or exotic probability theory. From the results presented above, one can see that while quantum amplitudes are used to compute probabilities, and while amplitudes and probabilities follow similar algebraic rules, quantum mechanics is not a generalized probability theory any more than information theory,[8] geometry, and number theory are generalized probability theories. Instead, they all derive from the same fundamental symmetries. Each is a specific measure theory in its own right.

A deeper understanding of the roles that mathematics plays in physics in terms of order, symmetries, and quantification will help to clear up mysteries related to the effectiveness and universality of mathematics as well as to guide future foundational efforts.

Acknowledgments I would like to thank John Skilling, David Hestenes, and Rob MacDuff for their insightful comments and ongoing discussions on foundations. I would also like to thank James Walsh for his careful proofreading, and Bertrand Carado and Yuchao Ma for their critical reading of the manuscript.

Technical Endnotes

This technical section provides the formal definitions of a partially-ordered set (poset) and a lattice as well as short derivation of the generalized sum rule for non-exclusive join.

A **partially-ordered set (poset)** P is a set of elements S along with a binary ordering relation, generically denoted $\leq$, postulated to have the following properties for elements $x, y, z \subseteq P$

For all $x, x \leq x$	Reflexivity
If $x \leq y$ and $y \leq x$, $x = y$	Antisymmetry
If $x \leq y$ and $y \leq z$, $x \leq z$	Transitivity

[8] The entropies appearing in the definition of mutual information derive from probabilities, yet no one insists that probability theory is an exotic form of information theory. Entropy and probability are related in a very specific way with probabilities being used to compute entropies. Similarly, quantum amplitudes are related to probabilities in a very specific way with amplitudes being used to compute probabilities via the Born Rule.

A poset $P = (S, \leq)$ is referred to as a **partially** ordered set since not all elements are assumed to be comparable. That is, there may exist elements $x, y \subseteq P$ where it is neither true that $x \leq y$ nor that $y \leq x$. In these situations, we say that x and y are **incomparable**, which is denoted $x \parallel y$.

A **lattice** L is a poset where each pair of elements has a supremum or least upper bound (LUB) called the **join**, and an infimum or greatest lower bound (GLB) called the **meet**. The meet of two elements $x, y \in L$ is denoted $x \wedge y$ and the join is denoted $x \vee y$. The meet and the join can be thought of as algebraic operators that take two lattice elements to a third lattice element. It is in this sense that every lattice is an algebra. The meet and join are assumed to obey the following relations

$$\begin{aligned} x \vee y &= y \vee x && \text{Commutativity} \\ x \wedge y &= y \wedge x && \\[1em] x \vee (y \vee z) &= (x \vee y) \vee z && \text{Associativity} \\ x \wedge (y \wedge z) &= (x \wedge y) \wedge z && \\[1em] x \vee (x \wedge y) &= x && \text{Absorption} \\ x \wedge (x \vee y) &= x && \end{aligned}$$

Lattice elements (and poset elements in general) can be quantified by assigning a real number (or more generally a set of real numbers) to each element. This is performed via a function v called a valuation, which takes each lattice element to a real number: $v : x \in L \to \mathbb{R}$.

Valuations are meant to encode the ordering of elements in the lattice and this is accomplished by insisting that for $x \leq y$ we have that $v(x) \leq v(y)$. Furthermore, if the valuation is to encode the relationships among the elements, it must be that the valuation $v(x \vee y)$ assigned to the join of two disjoint elements x and y (Fig. 3, left) can be expressed as a function of the valuations $v(x)$ and $v(y)$ assigned to those two elements. We write this as

$$v(x \vee y) = v(x) \oplus v(y).$$

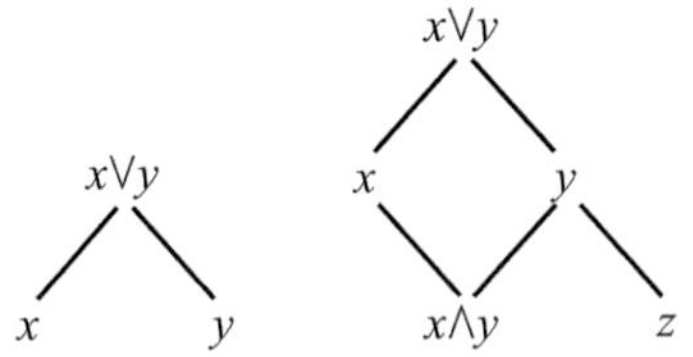

Fig. 3 (*Left*) An illustration of the join, $x \vee y$, of two disjoint elements x and y resulting in an additive measure. Disjoint elements have a null meet so that technically this structure is known as a join semi-lattice. (*Right*) The situation considered in the derivation of the Sum Rule, which applies to general cases

where the operator $\oplus$ is to be determined. The concept of a generally-useful quantification by valuation (non-degenerate ranking) implies that the operator $\oplus$ obeys a **cancellativity** property for disjoint elements x, y, and z, where for $v(x) \leq v(y) \leq v(z)$ we have $v(x) \oplus v(z) \leq v(y) \oplus v(z)$. Commutativity of the lattice join requires that the operator $\oplus$ is commutative, and associativity of the lattice join requires that the operator $\oplus$ is associative. The associative relationship represents a functional equation, known as the Associativity Equation, for the operator $\oplus$, whose solution is known to be an invertible transform of additivity [1, 4, 10], which can be written as

$$a \oplus b = f^{-1}(f(a) + f(b)),$$

where the function f is an arbitrary invertible function. In terms of the valuations this is

$$f(v(x) \oplus v(y)) = f(v(x)) + f(v(y))$$

which is

$$f(v(x \vee y)) = f(v(x)) + f(v(y)).$$

This suggests that one can always choose a simpler quantification than the valuations v by instead assigning values $u(x)$ defined by $u(x) = f(v(x))$ so that $\oplus$ transforms to simple addition for disjoint x and y:

$$u(x \vee y) = u(x) + u(y).$$

Recall that this result holds only for disjoint elements (in a join semi-lattice). We now derive the result for two lattice elements in general. Consider the elements $x \wedge y$ and z illustrated in Fig. 3 (right). Since their join is y, we have that

$$u(y) = u(x \wedge y) + u(z). \tag{A1}$$

Next consider that elements x and z are disjoint and their join is $x \vee y$. This allows us to write

$$u(x \vee y) = u(x) + u(z). \tag{A2}$$

Solving (A1) for $u(z)$ and substituting into (A2) we have the Sum Rule

$$u(x \vee y) = u(x) + u(y) - u(x \wedge y), \qquad \text{Sum Rule}$$

which holds for general elements x and y [8–10].

References

1. Aczél, J., Lectures on Functional Equations and Their Applications, Academic Press, New York, 1966.
2. Birkhoff, Lattice Theory, American Mathematical Society, Providence, 1967.
3. Cover, T. M., Thomas, J. A. 2012. Elements of information theory. John Wiley & Sons.
4. Craigen, R., Páles, Z. 1989. The associativity equation revisited. Aequationes Mathematicae, 37(2–3), 306–312.
5. Davey, B. A., and Priestley, H. A., Introduction to Lattices and Order, Cambridge Univ. Press, Cambridge, 2002.
6. Jaynes, E. T. 1959, Probability Theory in Science and Engineering, No. 4 in Colloquium Lectures in Pure and Applied Science, Socony-Mobil Oil Co. USA.
7. Klain, D. A., and Rota, G.-C., Introduction to Geometric Probability, Cambridge Univ. Press, Cambridge, 1997.
8. Knuth K.H. 2003. Deriving laws from ordering relations. In: G.J. Erickson, Y. Zhai (eds.), Bayesian Inference and Maximum Entropy Methods in Science and Engineering, Jackson Hole WY 2003, AIP Conference Proceedings 707, American Institute of Physics, Melville NY, pp. 204–235. arXiv:physics/0403031 [physics.data-an].
9. Knuth K.H. 2010. Information physics: The new frontier. P. Bessiere, J.-F. Bercher, A. Mohammad-Djafari (eds.) Bayesian Inference and Maximum Entropy Methods in Science and Engineering, Chamonix, France, 2010, AIP Conference Proceedings 1305, American Institute of Physics, Melville NY, 3–19. arXiv:1009.5161v1 [math-ph].
10. Knuth K.H., Skilling J. 2012. Foundations of Inference. *Axioms* 1:38–73, doi:10.3390/axioms1010038, arXiv:1008.4831 [math.PR].
11. Knuth K.H. 2014. Information-based physics: an observer-centric foundation. *Contemporary Physics*, 55(1), 12–32, (Invited Submission). doi:10.1080/00107514.2013.853426. arXiv:1310.1667 [quant-ph]
12. Knuth K.H., Bahreyni N. 2014. A potential foundation for emergent space-time, Journal of Mathematical Physics, 55, 112501. doi:10.1063/1.4899081, arXiv:1209.0881 [math-ph].
13. Knuth K.H. 2015. Information-based physics and the influence network. In: It from Bit or Bit from It? On Physics and Information, Springer Frontiers Collection, Springer-Verlag, Heidelberg pp. 65–78. FQXi 2013 Essay Contest Entry (Third Place Prize). arXiv:1308.3337 [quant-ph].
14. Goyal, P. 2015. Informational approach to the quantum symmetrization postulate. New Journal of Physics, 17(1), 013043. arXiv:1309.0478 [quant-ph].
15. Goyal P., Knuth K.H. 2011. Quantum theory and probability theory: their relationship and origin in symmetry, *Symmetry* 3(2):171–206. doi:10.3390/sym3020171
16. Goyal P., Knuth K.H., Skilling J. 2010. Origin of complex quantum amplitudes and Feynman's rules, Physical Review A 81, 022109. arXiv:0907.0909v3 [quant-ph].
17. Hamming, R. W. 1980. The unreasonable effectiveness of mathematics. American Mathematical Monthly, 81–90.
18. Pölya, G., and Szegö, G., Aufgaben und Lehrsätze aus der Analysis, 2 vols., Springer, Berlin, 1964, 3rd ed.
19. Sorkin, R.D. 1994. Quantum mechanics as quantum measure theory, *Mod. Phys. Lett.* A9 (1994) 3119–3128.
20. Wigner, E. 1960. The Unreasonable Effectiveness of Mathematics in the Natural Sciences, in Communications in Pure and Applied Mathematics, vol. 13, No. I (February 1960). New York: John Wiley & Sons Inc.

How Mathematics Meets the World

Tim Maudlin

No paper in the history of science has had a greater impact solely through its title than Eugene Wigner's "The Unreasonable Effectiveness of Mathematics in the Natural Sciences" [3]. The title conveys both a claim and a puzzle. The claim is that mathematics is the language in which our most accurate scientific theories are formulated. The puzzle is why the language of mathematics should be such an effective tool for describing the physical world. The claim is indisputable. The puzzlement requires some unpacking.

In its most radical form, the puzzlement could be directed at all of mathematics: why should any mathematical propositions have bearing on the behavior or structure of physical objects? Wigner himself does not raise this worry. Certain areas of mathematics struck him as unproblematic. He writes: "...whereas it is unquestionably true that the concepts of elementary mathematics and particularly elementary geometry were formulated to describe entities which are directly suggested by the actual world, ...". That mathematical concepts developed for the purpose of describing the physical world manage to do so reasonably well is not prima facie puzzling at all. Wigner's own conundrum appears in the completion of the cited sentence: "..., the same does not seem to be true of the more advanced concepts which play such an important role in physics". Wigner mentions complex numbers as an example of the more advanced concepts, and contemporary physics can supply more exotic examples.

We can therefore divide the initial question into two subquestions. (1) Which mathematical concepts seem naturally suited to describe features of the physical world, and what does their suitability imply about the physical world? (2) Why should any mathematical concepts that do not fall into the class of naturally suited ones nonetheless be of use in physics?

Among the naturally suited mathematical concepts are the integers. The usefulness of the integers for describing physical situations requires little on the side of

T. Maudlin (✉)
Department of Philosophy, New York University, 5 Washington Place, New York, NY 10003, USA
e-mail: tim.maudlin@gmail.com

A. Aguirre et al. (eds.), *Trick or Truth?*, The Frontiers Collection,
DOI 10.1007/978-3-319-27495-9_8

physics, but does require something. If there are physical items so constituted as to be solid objects, held together by strong internal forces and resistant to fracture and to amalgamation, then they will be effectively countable. A physical world completely described by fluid mechanics would contain no such objects, so the physics does make a crucial contribution. We can also imagine a physics whose fundamental constituents are perfectly discrete and "uncuttable"—the atoms of Democritus, for example—which are even better suited to unambiguous counting than are macroscopic solids such as tables and chairs. So the usefulness and limits of even the most basic mathematical concepts for describing the physical world are influenced by the physics. The only question is which physical features—such as practical solidity and indivisibility in this case—are relevant.

Even when the applicability of mathematics to the physical situation is straightforward, something akin to Wigner's puzzle may arise. A child playing with solid square tiles notices that sometimes a larger square array (5×5) can be rearranged into two smaller square arrays (3×3 and 4×4). This is a plain physical feature of the tiles. The purely mathematical fact that $3^2 + 4^2 = 5^2$ together with facts about plane geometry are relevant to explaining this. The child wonders whether the same situation ever occurs with his solid cubes: can a large cube made of smaller cubes be rearranged into two smaller cubes? This physical question can be resolved in the negative by proving Fermat's Last Theorem.

Fermat's Last Theorem was proven by Andrew Wiles as a corollary to the modularity theorem for semistable elliptic curves. Since Fermat's conjecture has to do with simple relations among integers, it might seem almost miraculous that any result about semistable elliptic curves would be relevant to it. And if one were to claim that a deep analysis of these curves would yield answers about the physical question of how cubes can be stacked, that might seem even more incredible. But the root of the surprise here is in not some uncanny connection between mathematics and physics, but rather in the uncanny bearing of one branch of mathematics on a seemingly unrelated distant branch. It may be astonishing that facts about semistable elliptic curves have implications about Fermat's Last Theorem, but it is not at all astonishing that the Theorem has implications for our cubes.

So one could easily write a companion paper to Wigner's called "The Unreasonable Relevance of Some Branches of Mathematics to Other Branches". Connections between apparently unrelated mathematical questions are regularly discovered, and any account of pure mathematics must acknowledge that. But we should not confuse this question with the entirely separate question of the effectiveness of mathematics in physics. If powerful and highly abstract mathematical methods are used to prove results in arithmetic or geometry and the relevance of the arithmetic or geometry for describing the physical world is clear, then Wigner's puzzle has been solved without remainder.

The relevance of the theory of integers for physics is unproblematic so long as the way that physical items are being counted is conceptually sharp. Different physical worlds and different physical concepts, as we have seen, can be more or less hospitable to unambiguous methods of counting. In the purely fluid world there might be little of interest to count. Even in the actual world, there can be a sharp answer to how many atoms there are in a particular strand of DNA, but no equivalently sharp answer to how many mountains there are in Europe. It is exactly for this reason that "mountain" is not a term that can be used in fundamental physics: mountains are not precisely enough delineated by physical conditions to be amenable to precise mathematical description. So the appropriateness of counting as a way to employ mathematics in describing fundamental physics turns on details of the fundamental physical ontology. If the world had turned out to contain eternal and physically indivisible Democritean atoms then there would be a precise mathematical characterization of how many atoms there are.

In this sense, Democritean atoms have a mathematical structure. Mathematical language can be used to provide answers to certain precisely articulated questions about them. But that is not to say that Democritean atoms are mathematical entities in any sense that would make them abstract or independent of the contingent physics of the world. In order to be precisely countable, the atoms must have some physical properties. Cells, whose exact numbers become indeterminate in the midst of cell division, do not always have those properties. So attributing a mathematical structure to physical items is not the same as postulating that they *are* mathematical entities: it is rather the weaker claim that they have some physical features that make them amenable to precise mathematical description in some respects.

We have sketched some of the physical properties required in order that things be describable unambiguously by the application of integers, i.e. that they be counted. And as soon as items are countable, other mathematical concepts can be brought to bear: ratios and proportions for example. Leopold Kronecker famously opined that "God made the integers, all else is the work of men". Although that is somewhat hyperbolic, there is a clear definitional route from the integers to the rational numbers, to the real numbers, to the complex numbers, etc. Insofar as the application of mathematics to the physical world runs through unambiguous counting, all of these more sophisticated mathematical items can be understood as more indirect mathematical tools for describing the physical structure of the world.

Wigner did not single out the integers as the mathematical entities most directly inspired by experience of the physical world, he mentioned "elementary geometry" instead. Wigner's position seems to be that there is no great puzzle about why the concepts of elementary geometry might be applicable to the physical world: those very concepts were developed via interaction with the world. If some early Greek counterpart to Wigner were to express puzzlement about the unreasonable effectiveness of geometry in the physical sciences (or the architectural or surveying sciences) he would be met with incomprehension. Geometry, he might be told, just is the study of points, lines, areas, shapes and volumes in space. Insofar as the architect or surveyor is interested in determining facts about such things, there is no mystery about the utility of geometry for the undertaking. Of course, physical objects are not perfectly suited for description using the vocabulary employed by Euclid. Plato insisted that the "square" and "rectangular" slabs of stone used by the architect are not square or rectangular in Euclid's sense, which is true. But even Plato did not deny that certain regions of physical space are precisely square, and that a physical stone could be a close approximation to such a shape. Just as Democritean atoms would have a precise arithmetical structure, so space itself was regarded as having a precise geometrical structure. If geometry is the study of that structure then it is hardly a surprise that it is useful for describing and interacting with the physical world.

The epistemic status of geometry was not well understood as long as it was taken to provide a priori and certain knowledge of the structure of physical space. This characterization naturally led to Kant's question of how any such a priori synthetic knowledge might be possible. The proper answer, of course, is that it is not possible: claims about the geometrical structure of physical space or space-time must be considered just as conjectural and fallible as all other physical claims. The development of non-Euclidean (and indeed non-Riemannian) geometries opened up a universe of geometrical possibilities. Choosing among them when framing physical hypotheses is an a posteriori and empirical matter, as it should be. So purely mathematical advances in abstract geometry due to Riemann and later Minkowski made the relation of geometry to physics all the more clear. Physical space (or space-time) has a geometrical structure. Various different abstract geometries correspond to different possible structures, and the only grounds for accepting one or another of these as accurate is the empirical success of a total physics that details both a space-time structure and the matter in it. It is exactly because there are alternative abstract geometries that there is less puzzlement about why one or another of them might accurately represent the physical structure of space-time.

But for geometry as for arithmetic, the foundational question remains. Numbers can be used to provide accurate and useful representations of physical situations insofar as those situations contain items that are unambiguously enumerable or countable. The unambiguity of a process of enumeration, in turn, depends on certain physical characteristics, such as the items having sharp boundaries and retaining structural integrity through time. Similarly, there should be some physical characteristics of a situation that make it amenable to geometrical description. It is in virtue of having these characteristics that the physical world can be usefully described by geometry; it is these characteristics that make the physical world into a geometrical object. But the question is: just what features must a physical entity have in order to display a geometrical structure?

The answer to this question for geometry is not obvious. It is nearly tautological to remark that enumerability is required in a domain in order to use numbers to describe it, and the further elaboration of characteristics that make objects unambiguously enumerable (the characteristics that Democritean atoms have but mountains do not) is not very contentious. No similar answer immediately presents itself for geometry.

In the remainder of this essay I will outline the standard answer to this question, and then argue that there is a better approach. The standard answer derives from the standard mathematical tool used to describe the most basic geometrical structure of a space. If that mathematical tool, that mathematical language, is to provide an accurate characterization of the geometry physical space or physical space-time then physical space-time must have a structure corresponding to the fundamental concept in the mathematics. And a different mathematical language, built on a different primitive concept, requires that physical space-time have a different structure if it is to be accurately described. I will argue that standard geometry has been *built on the wrong conceptual foundation* to apply optimally to space-time. I will sketch an alternative geometrical language, and explain how it could directly reflect the structure of the physical world.

There are various hierarchically organized levels of geometrical structure that a space can have. It can have a *metrical* structure, which describes the distance between points. This distance is just the minimal[1] length of a continuous path between the points. It can have an *affine* structure, which sorts continuous paths into straight and curved. It can have a *differentiable* structure, which distinguishes smooth curves from bent curves. But beneath all these, already presupposed by all of these, is the most basic geometrical structure: topological structure.

Attributing a topology to a space allows one to define basic notions of continuity. For example, in order characterize a function from a domain to a range as continuous, both the domain and the range must have a topology. So insofar as definitions of length or straightness or differentiability already presuppose the notion of a continuous path or a continuous curve they presuppose topological structure.

Carrying on our sifting humor, we may now ask: what features must a physical space or space-time have in order to have a topology? The fundamental concept of standard topology—the concept in terms of which all other topological notions are defined—is the *open set*. An open set of points is a set that has no boundary or edge, a set in which, intuitively, every point is completely surrounded by other points in the set. Specifying the topology of a space is exactly specifying the collection of open sets in the space. One then defines other geometrical features of the space in terms of the open-set structure. For example, a space is *connected* just in case it cannot be partitioned into two disjoint non-empty open sets.

If the mathematical language of standard topology turns out to be a powerful tool for describing physical space-time, then there is one obvious possible explanation: there is some *physical* property that imposes an open-set structure on the points of space-time. Just as the physical properties attributed to Democritean atoms would make them unambiguously enumerable and hence accurately describable using numbers, such an open-set-producing physical property would make physical space-time unambiguously describable using the conceptual resources of standard topology. All of the complex abstract results in topology would then hold of physical space-time, just as Fermat's Last Theorem (hard as it is to prove) settles the physical question about the possibility of rearranging physical cubes in certain ways.

The problem is that there is no obvious answer to the question: exactly what feature of the physical world invests space-time with an open-set structure? One can take this geometrical structure to be physically primitive and not subject to further explication. But that should be the response of last resort.

Perhaps the problem lies instead with the mathematical tool we are using. It is possible to develop alternative mathematical languages for describing the geometry of a space, based on different primitive notions than standard topology. For any such alternative mathematical language we can pose the same question: is there a feature

[1] Or, in Relativity, maximal.

of the physical world guaranteeing that physical space-time has the right sort of structure to be described using this language?

I have recently developed a mathematical language for describing the geometry of a space based on a different conceptual foundation than standard topology [2]. This language is called the *Theory of Linear Structures*, and the conceptual primitive in it, instead of the open set, is the *line*. In particular, the geometry of a space is specified by detailing which sets of points in it constitute open lines, "open" in the sense that the lines do not close back on themselves like a circle does. (See the appendix for the axioms of standard topology and the Theory of Linear Structures, as well as some technical results.)

What is geometrically characteristic of an open line is that it imposes a linear order on its points. In order to move continuously from one place to another on an open line one must pass through a unique set of intermediate points in a particular order. (On a circle, in contrast, on can move continuously from one place to another by two completely different routes: clockwise and counter-clockwise.) Furthermore, unlike open sets, lines have a strongly constrained geometrical relation between the part and the whole. Longer lines must, of necessity, have shorter lines as parts, whereas an open set need not have any parts that are themselves open sets. This permits us to lay down axioms for open lines that completely capture their geometrical nature.

The fundamental structural characteristic of an open line is this: given the points in an open line, there is a linear order among its points such that all and only the intervals of that linear order are themselves open lines. This basic structural characteristic of the open line holds for lines with infinitely many points (such as lines in Euclidean space) and lines with only finitely many elements (such as a line of movie-goers waiting to buy tickets). So the Theory of Linear Structures is capable of describing the geometry of continua and of discrete spaces (such as lattices) using the same conceptual and definitional resources. Standard topology does not fare well in this regard: intuitively, there are no non-trivial open sets in a discrete lattice, so the mathematical language of standard topology provides little geometrical information about it. In contrast, graph theory is a branch of the Theory of Linear Structures. But graph theory cannot deal with continua while the Theory of Linear Structures can.

There is a connection between the Theory of Linear Structures and standard topology worthy of notice. Once one has specified which sets of points in a space constitute lines, it is easy to define a *neighborhood* of a point p. A neighborhood of p is a set of points Σ such that every line with endpoint p contains a segment (i.e. a part which is itself a line) with endpoint p that lies in Σ. So if Σ is a neighborhood of p, then every continuous route to p must enter Σ before it arrives at p. This definition works equally well in a discrete space as in a continuum. Finally, we can define an *open set* in the space as a set of points that is a neighborhood of all its elements. The open sets, so defined, provably satisfy all of the axioms for open sets in standard topology. So specifying a Linear Structure for a space automatically, by this chain of definitions, specifies a standard topology for it. But the Linear Structure typically contains much more detailed geometrical information than the standard topology. In a connected discrete space like a lattice the only open sets are the entire space and the empty set, so the standard topology contains no detailed geometrical information. The Linear Structure of such a discrete space, in contrast, determines its entire graph-theoretical structure.

There is another notable difference between standard topology and the Theory of Linear Structures. If one were asked to impose a *direction* on an open set, it is not at all clear what is being asked for or how it can be provided. But to specify a direction on a line is perfectly straightforward. Lines in Euclidean space, for example, have exactly two directions. This corresponds to the fact that exactly two linear orders among the points yield the same set of intervals, namely the linear orders that are inverse to each other. So choosing a direction on a line is just choosing one of these two linear orders among the points. It is almost trivial, then, to extend the Theory of Linear Structure to include geometries with *directed* lines, lines that point one way rather than the other, like a one-way street. There is no similarly simple and straightforward way to introduce the notion of directedness into standard topology.

Returning to our original question: suppose instead of standard topology with its open sets we choose the Theory of Linear Structures as the mathematical language in which to describe the geometry of physical space or physical space-time. Then our fundamental question is this: what physical feature of the universe makes space-time amenable to description in this mathematical language? What physical feature of the universe might be responsible for creating *lines*, that is *linearly ordered sets of points* in space-time?

The choice between giving an account of the geometry of *space* (such as Euclid proposed) and given an account of the geometry of *space-time* (such as the General Theory of Relativity proposes) is now critical. The "points" of space-time are *events* that are perfectly localized in both space and time. So providing a geometry for space-time requires specifying a geometry of events. And asking for a source of lines in space-time is asking for a feature of physical reality that *linearly orders events*. But we all have always accepted that there is such a feature of physical reality, namely *time*.

Our understanding of the structure of time has been revolutionized by the Theory of Relativity. Intriguingly, the change from a classical to a Relativistic account of temporal structure is of exactly the right sort to promote time into the sole creator of physical geometry. In a classical space-time, such as Newton imagined, each "moment of time" is a global sort of thing: events all over the universe take place "at the same time". In other words, a classical space-time structure has an absolute and universal notion of simultaneity of events. We can ask for a set of events in a classical space-time that are linearly ordered by time (so no two events in the set are simultaneous). We can even ask for a *maximal* such set, a set to which no more events can be added while retaining the unique linear order in time. But in a classical setting, such maximal sets of linearly time-ordered events are typically not continuous lines in space-time. In a classical setting, such maximal time-ordered sets of events are typically more like a random scatter: one event at each moment of time (to make the set maximal), but with otherwise no relation among the events.

In Relativity, though, things work out quite differently. The difference arises because the realm of events with no defined time-order relative to a given event p is much larger in relativity. The only events that are time-ordered with respect to

p are events in or on either the future or past light-cone of p. So any totally time-ordered set of events that contains p contains no events at space-like separation from p, and for the same reason contains no events space-like separated from any event in the set. This creates a much, much stronger constraint on maximal time-ordered sets of events in Relativity than the corresponding constraint in a classical space-time setting. Indeed, maximal sets of time-ordered events in a (globally hyperbolic) space-time correspond to the continuous timelike-or-lightlike trajectories through space-time. That is, referring to nothing but the linear temporal ordering of events in a Relativistic space-time, it is easy to define sets of events that we intuitively regard as continuous lines in space-time. Indeed, they are intuitively continuous *directed* lines in space-time, where the directionality is the direction of time.

So if one asks, using this mathematical language, what physical feature of the universe creates its geometry and makes it amenable to mathematical representation, that becomes the question: what physical feature imposes a linear order on events that defines the continuous lines in the physical arena? And in a Relativistic (but not classical) setting, a perfectly acceptable answer is: that physical feature is time. Time creates geometry and makes gives the universe a structure described by geometry: geometry as articulated using the Theory of Linear Structures.

It is interesting to consider how much geometrical structure is already built into space-time at the most fundamental level according to this application of the Theory of Linear Structures. If the lines in space-time are the timelike-or-lightlike lines, then just from this set of lines one can easily recover the entire light-cone (conformal) structure of space-time. The future light-cone (including its interior) of p is just the set of points one can reach from p going along a line forward in time, and the past light-cone the set of points one can get to going backward in time along a line. The lightlike geodesics can also be easily characterized: a line with endpoints p and q is a null geodesic if there is only one line that has those events as endpoints.[2] So quite a bit (although not all) of the Relativistic geometry of a space-time is already built in at this level of geometrical description.

By way of contrast, in standard topology nothing characteristically Relativistic is built in at the most fundamental level of geometrical description. In the standard approach, the topology of a Relativistic space-time is just a four-dimensional manifold, exactly the same topology of a classical space-time or even a four-dimensional Euclidean space that has no temporal aspect at all.

It is also notable that since the Theory of Linear Structures applies with equal ease and facility to discrete spaces and continua, there is no conceptual or technical barrier to using it to construct discrete Relativistic space-times. One advantage to a discrete space is that it comes automatically equipped with a measure—counting measure—that can be used to define the sizes of lengths of things. In a continuum, every finite magnitude contains infinitely many points, so counting is no guide to size. In a discrete case, counting yields non-trivial results, and the *entire* geometry (including metrical structure) can be built in at this most fundamental level in a

[2]Turning the "if" in this characterization into "if and only if" requires dealing with conjugate points, which can easily be done using the constructive techniques mentioned in the Appendix.

discrete Relativistic space-time. One does, however, have to be careful about just what to count. This will serve as our final illustration of how the resources of this mathematical language can be used.

If the most basic geometrical structures in a space are lines, then the most basic metrical structure will attribute lengths to the lines. In the context of a discrete geometry, the first thought is to quantify the length of a line by counting nodes or points in it. A line made up of 8 points is twice as long as a line made up of 4, etc.

But the metrical structure of a Relativistic space-time is conspicuously unlike the metrical structure of Euclidean space or any Riemannian curved space. The "proper length" of a lightlike geodesic is always zero. (This sometimes leads to the odd assertion that "time does not pass for a photon", even though the photon may be involved in a sequence of interactions that is ordered in time: first passing though one filter and then through another, for example). A Relativistic geometry is not, in the formal sense, a "metric" at all: the "lengths" do not obey the Triangle Inequality. So we should already be on guard that the account of "lengths" in space-time will not be exactly what one might expect.

Presenting familiar Relativistic space-times in terms of their Linear Structure already provides some clues. As we have said, there is no need for the existence of any "spacelike" lines at all: the whole geometry is coded up in the structure of the timelike-or-null trajectories. And among these, the null (lightlike) trajectories are distinguished: if there is exactly one line with a given pair of events as endpoints then it is a null trajectory, and if there is more than one then both trajectories are timelike. (This criterion can be restricted to suitably small regions to avoid long-distance counterexamples.) So if one carries this very same criterion over to a discrete setting, all minimal two-point lines will automatically be null lines. Any discrete Relativistic space-time is built out of fundamentally lightlike smallest parts.

One can also provide a criterion for a trajectory in a discrete space-time to be "bent" at a given node. Consider the three-point trajectory *a-b-c*. If this is the only line with *a* and *c* as endpoints, then we will say the line is *unbent*, and if there is another such line (*a-d-c*) then the line is *bent* and *b* is a *corner*. If *a-b-c* is unbent then *a* and *c* are lightlike related and if it is bent then *a* and *c* are timelike related. The trick is now to attribute length to the lines not by counting *nodes* but rather by counting *corners*. It immediately follows, by this definition, that lightlike geodesics have zero "length", just as the standard Relativistic (pseudo-)metric requires. The more corners a trajectory has—the more it "zig-zags"—the longer it is. Finally, we define the "distance" between any two events as the length of the *longest* line connecting them. This yields the "anti-Triangle Inequality" that is characteristic of a Relativistic geometry.

In this discrete Relativistic setting all of the geometrical structure, including the metrical structure, gets built in at the most fundamental geometrical level. Furthermore, since the smallest parts of the geometry—the simple two-point lines—are automatically lightlike one would be forced to the conclusion that massive particles are, at a microscopic level, actually following highly bent, zig-zag paths. This is a suggestion that has recurred throughout the history of quantum mechanics, usually discussed under the name *Zitterbewgung*. But here we have reached the limits of our speculations. Our topic has been the geometrical structure of space-time and the correct mathematical language to describe it. Adding material contents to that space-time is the next step, where quantum-mechanical considerations come into play.

Wigner's question was this: why is the language of mathematics so well suited to describe the physical world? A proper answer to this question must approach it from both directions: the direction of the mathematical language and the direction of the structure of the physical world being represented. In order for the language to fit the object in a useful way the two sides have to mesh. The fundamental concepts of the mathematical language—whether enumeration or the open set or the line—should have clear and unambiguous application to the physical world, and that can only obtain if the world has the right kind of structure. What form that structure takes depends on the mathematical language being used.

Physicists seeking such a mesh between mathematics and physics can only alter one side of the equation. The physical world is as it is, and will not change at our command. But we can change the mathematical language used to formulate physics, and we can even seek to construct new mathematical languages that are better suited to represent the physical structure of the world. The Theory of Linear Structures, whatever else its virtues, provides an example of how this can be done. If it is correct, then we might see how the time itself creates the geometry of space-time, and also makes space-time exactly the sort of thing that is well described using this mathematical language.

Appendix

These are the axioms of standard topology as presented in Crossley [1]:

> Definition: A **topological space** is a set, X, together with a collection of subsets of X, called "open" sets, which satisfy the following rules:
> T1. The set X itself is "open".
> T2. The empty set is "open".
> T3. Arbitrary unions of "open" sets are "open".
> T4. Finite intersections of "open" sets are "open".

There are four flavors of Linear Structure, depending on whether the lines have no directions or have directions, and whether conjoining lines requires that they share only an endpoint (point-spliced) or overlap by a segment (segment-spliced). Terms in bold are proprietary to the Theory of Linear Structures. A **segment** of a **line** λ is a subset of the points in λ that is itself a **line**. A **closed line** is a **line** with two **endpoints**.

Recall that a linear order $\geq$ on a set is a total relation that is antisymmetric and transitive. An interval of a linearly ordered set is a subset Σ such that if $p, q \in \Sigma$ and $p \geq r \geq q$, then $r \in \Sigma$. An endpoint of a linearly ordered set is an element that does not lie between any other pair of elements.

The axioms for an undirected point-spliced Linear Structure are:

A *Point-Spliced Linear Structure* is a set S together with a collection of subsets Λ (called the **lines** in S) that satisfy the following rules:
LS_1 (Minimality Axiom): Each **line** contains at least two points.
LS_2 (Segment Axiom): Every **line** λ admits of a linear order among its points such that a subset of λ is itself a **line** if and only if it is an interval of that linear order.
LS_3 (Point-Splicing Axiom): If λ and μ are **lines** that have in common only a single point p that is an **endpoint** of both, then $\lambda \cup \mu$ is a **line** provided that no **lines** in the set $(\lambda \cup \mu) - p$ have a point in λ and a point in μ.
LS_4 (Completion Axiom): Every set of points σ that admits of a linear order $\geq$ such the **closed lines** in σ are all and only the closed intervals of $\geq$ is a **line**.

In a sense, all of the fundamental work in the definition is done by the Segment Axiom. Any collection of subsets that satisfies only the Segment Axiom (which we call a *Proto-Linear Structure*) can be extended in a unique way to satisfy all four axioms.

To define a *Directed Linear Structure* we require that each **line** have a direction, i.e. be represented by a unique linear order rather than (as with undirected **lines**) a pair of linear orders. An **endpoint** of a **directed line** is therefore either an **initial endpoint** or a **final endpoint**. The Point-Splicing Axiom is adjusted so that the point in common must be the **initial endpoint** of one directed line and the **final endpoint** of the other.

The definitions of a **neighborhood** and an **open set** in an undirected Linear Structure are as follows:

A set σ is a **neighborhood** of a point p iff σ contains p and every **line** with p as an **endpoint** has a **segment** with p as an **endpoint** in σ.
A set σ in a Linear Structure is an **open set** iff it is a **neighborhood** of all of its members.

In a Directed Linear Structure we can distinguish **inward** and **outward neighborhoods** of p depending on whether p is required to be the **initial** or **final endpoint** of the **lines**. We can therefore define **inward** and **outward open sets**.

It is a theorem that the **open sets** (or **inward** or **outward open sets**) of a Linear Structure satisfy the axioms T1–T4 above. So investing a set of points with a Linear Structure automatically invests it with a standard topology. However, if one uses only undirected Linear Structures, then many standard topologies cannot be generated this way. For example, given a set of 5 points, there are 6,942 distinct topologies that can be put on the set, but only 1,024 distinct Linear Structures. These Linear Structures, in turn, generate only 52 distinct topologies. So only 52 out of 6,942 standard topologies can be recovered. However, if one uses Directed Linear Structures the situation changes dramatically. There are 1,048,576 distinct Directed Linear Structures that can be put on the 5 points, and these generate all 6,942 possible standard topologies. Many distinct Directed Linear Structures generate the same topology, so the Directed Linear Structure contains more geometrical information than the standard topology.

This allows us to make a distinction among standard topologies that has not been made before (to my knowledge). A topology is **intrinsically directed** iff it can be generated from a Directed Linear Structure on the points but not from any undirected Linear Structure. We can prove another theorem: Every finite-point topology can be generated from some Directed Linear Structure. This theorem does not hold, however, if there are infinitely many points. There are some topologies in that case which provably cannot be generated by any Directed Linear Structure on the points. We call these **geometrically uninterpretable** topologies.

This provides a sample of what can be defined using the Theory of Linear Structures and some of the theorems that have been proven.

References

1. Crossley, Martin (2005) *Essential Topology*. Dordrecht: Springer
2. Maudlin, Tim (2014) *New Foundations for Physical Geometry: The Theory of Linear Structures*. Oxford: Oxford University Press
3. Wigner, Eugene (1960) "The Unreasonable Effectiveness of Mathematics in the Natural Sciences". *Comm. in Pure and Applied Math.*, Vol. 13, No. 1

Mathematics: Intuition's Consistency Check

Ken Wharton

Abstract There is a well-noted overlap between mathematics and physics, and in many cases the relevant mathematics was developed without any thought of the eventual physical application. This essay argues that this is not a coincidental mystery, but naturally follows from (1) a self-consistency requirement for physical models, and (2) innate physical intuitions that guide us in the wrong directions, slowing the development of physical models more so than the related mathematics. A detailed example (concerning the flow of time in physical theories) demonstrates key parts of this argument.

Introduction

Surveying modern physics, one sees higher mathematics everywhere. Rare and esoteric branches of math have been put to use by physical models, often long after the math was first developed. Indeed, so many different branches of mathematics have found effective physical applications that it raises some deep questions [1].

Perhaps the most interesting of these questions generally concern the tight overlap between higher mathematics and our successful physical theories. All of our physics utilizes mathematics, and we can barely find any mathematical topics that have not been put to some physical application [2]. This close relationship calls out for some explanation. Given that physics and math are such different pursuits, why is there so much convergence?

This essay will argue that there is no real mystery here at all. Yes, physics successfully *uses* higher mathematics to make predictions about the actual workings of our universe, and the mathematics was not necessarily developed with such applications

K. Wharton (✉)
Department of Physics and Astronomy, San José State University,
San José, CA 95192-0106, USA
e-mail: kenneth.wharton@sjsu.edu

A. Aguirre et al. (eds.), *Trick or Truth?*, The Frontiers Collection,
DOI 10.1007/978-3-319-27495-9_9

in mind.[1] But to see this as a mysterious convergence is to ignore a crucial fact: "Physics" is a tiny slice of somewhat-successful models winnowed out of a vast heap of failed and inconsistent ideas.

The next section will explain how this winnowing process naturally selects for physical models that can be framed in the language of mathematics. Section Intuition Versus Objectivity tackles the other key mystery, arguing why it's reasonable that *most* mathematics will get incorporated by one physical model or another.

One crucial assumption in these arguments has to do with physical ideas that fail in a particular manner. The following section is therefore a deep-dive into one such example, meant to illustrate this central point (and incidentally, also explaining the pervasive use of the "block universe" in physical theories). Conclusions follow.

The Deduction Factor

The apparent mystery of the physics-mathematics connection stems, in part, from the observation that these are quite different pursuits. But while these differences are important (and will explain why the math often comes first) there are also essential similarities.

Whatever the motivation, mathematics formally begins with axioms: a set of rules that are correct by definition. Mathematicians take great pains to ensure that any set of mathematical axioms is self-consistent; following the axioms to their logical conclusions, one should never encounter a contradiction. Starting with only the axioms, one can deduce various theorems, extending the domain of mathematical concepts and consequences.

One analogy (albeit imperfect) is that self-consistent axioms are like cave entrances, and the deductive process is an exploration of the caves. This exploration reveals which statements are supported by the axioms (the caves) and which ones are not (the solid rock). There are other important aspects of mathematics—building higher-level concepts that can be explored in their own right—but for our purposes this is just like finding a new set of caves, deeper in.

Obviously, fundamental physics is different. Our deepest physical theories aren't treated as axioms, but are instead hypothesized to explain observations. Coming up with fundamental physical laws is therefore an *inductive* process. Furthermore, physical theories have to contend with experimental observations that may shed doubt on entire lines of established work. Mathematicians do not have such concerns.

But focusing on these differences between physics and math obscures an important similarity: physics uses the deductive process, too. *Once* the hypotheses are generated, physicists use deduction to determine the predicted consequences. (These consequences are then compared to experimental observations.) This part of physics is just like math; the hypotheses are entrances to caves, and physicists explore the

[1]Although of course mathematicians are influenced (at least in part) by scientific problems and progress, and many individuals have worn both physicist- and mathematician-hats.

space of consequences just like mathematicians. Sure, physicists are often chased out of caves by the monsters of experimental falsification, but the deductive exploration is still essentially the same.

It is this crucial similarity that explains why physicists use mathematics. There's no point in pursuing hypotheses that are internally inconsistent; if a hypothesis leads to self-contradictory implications, it's ruled out before the first experimental test. The only caves that physicists find worth entering are the structurally-sound ones that have been mapped out by the mathematicians. Translating physical theories onto some coherent mathematical structure is therefore a necessary consistency check.

If you can't translate a physical theory into known mathematics, it's either fatally inconsistent or a harbinger of a new mathematical field. But the latter is very unlikely. Part of the reason that mathematicians have (often) mapped out the entrances to new caves before physicists arrive is that they can be more fearless. No monsters of experimental falsification lie in wait, and there's no assumption that there's one "correct" cave to be exploring.

It's important to note that the above argument concerns *mathematical* consistency, not conceptual consistency. Blurring inconsistent concepts together has arguably been the source of several fruitful advances, but that's different from working with concepts that can't even be expressed via mathematics in the first place. Physics without mathematics would be a surefire route to failure.

Intuition Versus Objectivity

Using mathematics as a consistency check may sound reasonable, but why does physics successfully use such strange and esoteric mathematics as differential geometry and Clifford algebras? Just because a given field of mathematics is consistent doesn't mean it should actually be useful. These connections might be thought to imply that abstract math is somehow hitting upon secrets of the universe long before the physicists even realize it.

But a more likely explanation is that math is just less *biased* than physics, more willing to explore new ideas. In both fields, the cave entrances still need to be invented by people: mathematical axioms and physical hypotheses have to be dreamed up in the first place. And it's perfectly reasonable that the most intuitive axioms/hypotheses are going to be dreamed up more often, with less intuitive ideas being much rarer.

Furthermore, it should be evident that we have a lot more innate intuitions about physics than we do about math. (Not that we have more *correct* intuitions; just more intuitions in general.) Every day an average person experiences a vast array of physical phenomena, but only very minimal mathematics. Evolution has taken these physical experiences and found that some ways of parsing our physical world are more useful than others—effectively giving us innate, instinctual biases about how physical things "work".

These intuitions are therefore a stronger influence on physicists than on mathematicians. The hypotheses that a given physicist finds worth pursuing are at least partially constrained by the intuitions of that individual. But with less hardwired intuitions about anything beyond, say, fractions and integers, mathematicians come at their field with less evolutionary baggage.

Of course, professional physicists and mathematicians have gained non-instinctive intuitions, in the course of mapping out known caves and seeing what others have discovered. But again, these intuitions are more likely to constrain physics than math. No one is trying to find one "correct" set of mathematical axioms; mathematicians can explore some totally different cave entrance without claiming all the other caves are wrong. Physicists, on the other hand, gain intuitions about which hypotheses to *avoid*, and perhaps even board up the entrances to unpromising caves with unexplored territory remaining inside.

Unencumbered by such physical intuitions, it seems reasonable that mathematicians are simply better explorers of new and strange ideas. Granted, mathematicians are more likely to go deeper into a particular cave once the physicists see a potential application, but the cave *entrances* (the axioms) are often mapped out before this happens. And while experimental monsters play a role in forcing physics out of comfortable caves, pushing the creation of new physics ideas, those experiments first have to *happen*. That can take a long time, especially if the relevant physical scales are difficult for us humans to access (the very phenomena for which our intuitions have had no evolutionary guidance). Mathematicians aren't waiting around for any such experimentally-imposed revolutions; they're out exploring new cave entrances while the physicists are settling into the old ones.

Figure 1 shows a rough cartoon of how these biases might impact the space of ideas about our physical universe. The vertical axis separates ideas that can be mapped onto some set of mathematically-consistent axioms (at the bottom), from ideas that

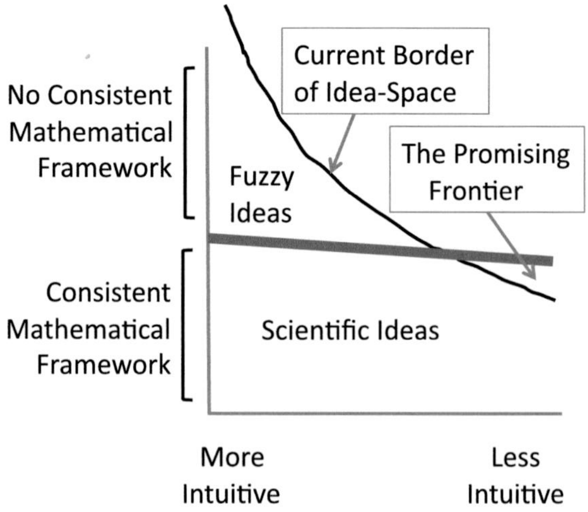

Fig. 1 A cartoon of idea-space about the physical world, separated into domains of mathematical consistency (*vertical*) and intuitive plausibility (*horizontal*)

can't.[2] The horizontal axis separates idea-space on a gradient of how "intuitive" the idea might seem; almost by definition, we have come up with more intuitive ideas than counter-intuitive ones.

There are two key debatable features of this cartoon. First, it indicates that many of the perfectly-intuitive ideas people have had about physics are unscientifically "fuzzy", in that they do not map onto a consistent mathematical framework. Second, it indicates that the "Promising Frontier" (the space where new successful ideas are more likely to be produced) lies on the non-intuitive side of idea-space. Setting the first issue aside until a later section, an objection to this second feature could be raised if there are objectively *many more* intuitive ideas than counter-intuitive ideas. This might then move the Frontier back towards the (perhaps much larger) space of unexplored intuitive ideas.

Such an objection might be reasonable for some physical sciences, but to judge by the past century, it does not seem reasonable for fundamental physics itself. Recent progress in our deepest understanding of fundamental physics has relied on quite esoteric mathematics, just where the Frontier is located in Fig. 1. Objections along these lines might be "intuitive", but that is precisely the problem: it conflates what seems natural to us humans with some objective measure of naturalness. The fact that our most successful fundamental physics continues to be so *counter*-intuitive sheds doubt on this objection.

With all this in mind, reconsider the question posed at the start of this section: Why does physics use such strange and esoteric mathematics? The key words here are "strange" and "esoteric". Those are subjective delineations, not objective ones. If we suppose that mathematics has brushed those subjective categories aside and mapped out some reasonable *objective* distribution of self-consistent caves, then the fact that all these caves are finding uses in the physical sciences is simply an indication that our physical intuition is generally and profoundly mistaken. What we might have taken as a useful clustering of hypotheses on the "intuitive" side of Fig. 1, has turned out not to be useful after all.

Furthermore, there is not just one type of physical problem looking for one type of mathematics to explain it. Physics describes a vast array of phenomena at different scales, under many different approximations. Given this large number of very different problems, it's no surprise that a lot of mathematics gets used. To take one very simple example, states in classical statistical mechanics are not three-dimensional like the space of our universe, but instead naturally live in a 3N-dimensional "configuration space". (N is the number of particles.) The fact that the mathematics of such configuration spaces is useful is not an indication of how many dimensions we "really" live in, but simply a reflection of what kind of math is most useful to us when we don't know the microscopic details of a complicated system. In general, we come at problems from so many different perspectives that it's easy to see why we might use many different mathematical fields in our physical theories.

[2] There's always the possibility that some new mathematical system will be developed that will push the thick separation line upwards, but as argued here, it's more likely this will have already happened before the relevant physical idea comes along.

This argument is remarkably robust. Perhaps our intuition should be given more credit, and there should be more self-consistent axioms found on the left of Fig. 1. (Say, a steeper thick line.) Or perhaps mathematicians are more blinded by intuitions than one might think, and there's a vast array of unexplored consistent math on the right of Fig. 1. But the above argument only requires that our physical hypotheses are *more* biased towards incorrect intuitions than are mathematical axioms (and also that our intuitions are at least somewhat misleading).

If this argument goes through, the mystery here is not a deep mystery about the universe, but a human one: why are our physical intuitions so wrong at the deepest level? The fact that at least some of our intuitions *are* wrong has been well-established for the last century; if we take it as a given, then the original mystery seems to be essentially solved.

Fuzzy Ideas: An Explicit Example

The argument presented in the previous section posits that physics ends up using "strange" mathematics because our innate intuitions about the physical world are so wrong, at the deepest level. As our incorrect physical theories get ruled out, we're forced into the self-consistent (but non-intuitive) domain of higher-level abstract mathematics. This domain also took longer for mathematicians to develop, but they generally got there faster than the physicists.

If this argument is correct, the pressure towards "innately intuitive" ideas should be evident, even today. After all, if physicists explored all ideas equally, intuitive or not, the current border of idea space indicated in Fig. 1 would flatten out. The promising frontier would no longer lie in the domain of abstract mathematics, and physicists would be just as likely as mathematicians to stumble upon some new self-consistent framework.

The above arguments also noted that the use of mathematics in physical theories is not a surprise coincidence, so long as there exists an idea space that does *not* have a solid mathematical representation. Once these inconsistent "fuzzy ideas" are eliminated, the remaining "scientific ideas" use the language of mathematics to make consistent and objective predictions.

A reasonable objection to the above arguments is to deny that the space of "fuzzy ideas" even exists, or that the ideas in that space are ever seriously pursued by any physicist. If such ideas did not exist, there would be no winnowing process to explain the convergence of physics and math.

The previous sections would therefore be bolstered by an explicit discussion of a class of ideas which (1) seem very intuitive, (2) continue to be seriously considered by physicists, and (3) do not map onto any consistent mathematical framework. It is this sort of idea, which, after being winnowed out on the grounds of mathematical failure, explains the physics-math convergence. Furthermore, the appeal of such ideas on the intuitive-side of Fig. 1 helps to explain why there continues to be a "promising frontier" on the non-intuitive side. So with that in mind, let us turn to the inexorable flow of time.

The Intuitive Flow of Time

Whether or not 'time flies', it's certainly intuitively obvious that it 'flows'. Right now, the future "lies ahead"; once it has "passed" it will be "behind us". This relative temporal "motion" is surely some sort of "flow". Without those quotation marks it would be easy to read such statements without noticing that they naturally treat time as a *spatial* dimension. This natural metaphor appears in every language, although in China the future tends to be "down" and there is an Andean language in which the future is behind us [3].

Whatever the relative orientation, it certainly seems reasonable that if a particular event (at one particular clock time) is moving around in some analog to space, then it must be experiencing some analog to flow. Sure, there's a nagging concern that perhaps such a sentiment is unfairly conflating our subjective *perception* of time with time itself, but let's ignore that issue for the moment, or at least imagine that our perceptions are telling us something about time that is objectively true.

A ubiquitous analogy is that time is like a river. The future is obviously downstream, because the water of time is always flowing in that direction. After all, we say that "the time will come" and that "deadlines are approaching"...Oh, wait; that must mean the future is *upstream*, coming towards us: "The future is upon us", and "the past is receding". Or instead, maybe it's *us* that's moving relative to time; "we're coming up on the deadline" and then "we passed the deadline". These incompatible metaphors [3] are another potential cause for concern, but we can still generally agree that *something* is flowing, that there is some literal sense in which time "passes".

But if you try to look for this perfectly-obvious flow of time in our standard descriptions of physical phenomena, you won't find it. The trajectory of a baseball can be parameterized as $x(t)$, but this is just a static curve telling us the location (x) at any time (t). Draw this curve on a graph, and the line just sits there on the page, not going in either direction. Describe the electric field of the universe in (3 + 1D) spacetime, $\vec{E}(x, y, z, t)$, and you'll get a 4-dimensional function that also doesn't "flow" in either direction.

Of course, you can use such a function to make a movie of $\vec{E}(x, y, z)$, and play it forward, and it may show electromagnetic waves going in one direction. But you can also play it backward, and you'll see them going in the other direction. Physics doesn't tell us which way to play the movie (and indeed, does not even change the equations when run backward). Given a complete description of the classical universe, the direction that we play the movie is still something we have to put in by hand.

Depending on exactly how one interprets quantum theory, one might come to a slightly different conclusion as to whether there is a "flow" of time hiding in that particular formulation. (Although the following analysis applies to quantum theory as well.[3]) Regardless, we experience the flow of time on a macroscopic level, and at

[3]Furthermore, the phenomena described by quantum theory are far more obviously time-*symmetric* than everyday phenomena, making this an odd place to suddenly give time a directional flow.

the level that describes observable large-scale phenomena, any flow is absent from the relevant descriptions.

To some (here, unnamed) physicists and philosophers, this failure of physics to describe the flow of time has been viewed as a serious shortcoming. Yet, despite various efforts to fix this situation, there is still essentially no physical theory that matches our intuitions when it comes to the passage of time. The reason comes down to mathematics.

The Two-Time Trap

The mathematical template for ordinary "flow" or motion is clear enough. Consider again the baseball trajectory represented as $x(t)$. If x increases with increasing t, then we say the ball is moving in the positive-x direction. Even in a continuous medium, one can represent the flow of a wave in terms of the displacement D of that medium over space and time; $D(x, t)$. For a wave structure that maintains its shape as it moves with a velocity v in the x-direction, one is looking for solutions of the special form $D(q)$, where $q = x - vt$.

But a straightforward extension of this "flow through space" mathematics to the concept of "flow through time" requires that one turns space into time. And a naive replacement of $x \rightarrow t$ hits a serious, immediate problem. The baseball analogy would lead one to the non-function $t(t)$, and the wave analogy begs the question of how "fast" time flows. The only plausible answer to this question is a "velocity" of one second per second, at which point q simply turns into zero.

The problem here is that "time" is already built into *ordinary* flow, and extending the concept to time itself immediately runs into dangerous and circular logic. This same problem was evident in the above river example; if time is simply the physical distance along the river, what could it even mean to say that the river is "moving"? Clearly, the river analogy requires that one keeps some semblance of "ordinary time" still mentally running in the background, even as one maps some other semblance of time to a spatial parameter. This example, and many like it, suffer from having two completely separate notions of time, considered together in some confusing mishmash—perhaps explaining our tendency to use incompatible metaphors.

Attempting to carefully map our intuitions onto a mathematical framework therefore requires us to separate these different notions of time, perhaps looking for something like t(T), where t is *one* kind of time, and T is *another*. This is a plausible path to pursue, but it requires a fateful choice. Either:

(A) Both t and T map to ordinary clock time.

(B) Only t is ordinary time; T is not.

(C) Only T is ordinary time; t is not.

(D) Neither t nor T are ordinary time.

But option (A) immediately runs into the same problems that we saw for $x \rightarrow t$. If both of these quantities can be mapped to clock time, then t and T are (to within some known transformation), the very same thing. Such maps would force the use of some particular function $t(T)$ by definition, even before any particular behavior can be described with it. Therefore this approach can't describe any new phenomena, such as a purported flow of time. True flow is a *relative* statement; say, a distance changing with respect to time. Unless t and T are essentially different, measuring one with the other is just a meaningless tautology.

Option (B) runs into different issues. Perhaps our entire universe (including t!) is just a computer simulation, and T is the clock time of the master computer. But then T is only meaningful from the perspective of the computer; it should mean nothing to us. In fact, comparing our original intuition of flow from $x(t)$ to some new $t(T)$, it would mean that our intuition of the flow of time is absolutely and completely mistaken. Time doesn't flow, it "flerbles", a concept that is only definable in terms of some new concept T. Whatever this new concept, it needs to be incorporated into the rest of physics in some coherent manner, without drifting back into option (A). And therefore "flerble" cannot in any way mean "change", for change is with respect to ordinary time, and T is not ordinary time.

Next consider (C), which at least salvages the notion of change. But the entity that's changing now doesn't map to ordinary time. Something is flowing, perhaps, but not time. And (D) suffers from both the shortcomings of (B) and (C), without the advantages of either. Even if ordinary time somehow emerges at some higher level from t or T, it wouldn't be able to emerge from *both* without ending up back at option (A).

Especially when it comes to option (B), these conclusions can be obscured using linguistic sleight-of-hand. Instead of "change" or "flow", someone might propose that time is instead "becoming", "actualizing", or "condensing from potentiality". Maybe there are verbally-expressible notions of change-without-time, or flow-without-change. Or perhaps everything is best framed in terms of meta-time, super-time, timeless-time. The idea space of how to address this issue is literally boundless.

But eventually, if these ideas are going to make it out of the "fuzzy idea" space of Fig. 1 and into the domain of "scientific ideas", they must be mapped onto some particular mathematical framework. It is this process that lays the linguistic tricks bare; either there is a coherent mathematical framework, or there isn't.

Given the above objections, the only plausible path forward that might rescue our intuitions would be if there are indeed two different time dimensions (t and T) and we have some access/experience to *both* of these independent entities. Then it might be fair to claim that t and T might be subjectively experienced as "time" in different ways; t-time might be T-changing. But this step is essentially a hypothesis that we live in a universe of three spatial dimensions and *two* time dimensions (3 + 2D, not the usual 3 + 1D). And this is not a hypothesis with trivial consequences, to say the least. It's akin to inventing an entirely new set of mathematical axioms, with new caves of consequences to explore, and essentially none of it has yet to be connected with anything we know about our universe.[4] There may be interesting future physics in this direction, but it is not some gentle tweak to what we already know.

[4]There are a few interesting efforts in 3 + 2D spacetime, but not many [4].

Since this last remaining path is *not* generally pursued, we're left with physics that treats time as a directionless parameter—a "block universe" mathematical representation. Time is just a dimension; it does not "go" forward any more than space "goes" left. Caveats about future mathematical discoveries aside,[5] the requirement of internal self-consistency forces physics into a less-intuitive direction, a framework in which there is simply no objective flow of time.

Conclusions

We have lots of ideas, but only some of them can be mapped onto self-consistent mathematics. And self-consistency is a prerequisite for experimental testing; if a model predicts two contradictory observations, there is nothing to test. Physics and math therefore share the same essential need for internal consistency, and this makes mathematics the first step on the road to a viable physical theory.

The previous section demonstrated an example of a plausible idea that has not succeeded in making this first step. There are other examples of "obvious" ideas with no mathematical implementation, often revolving around our intuitions of time. Not only is there no flow of time, but there is no indication of the special instant "now" in our physical theories, and there seems to be no objective (agency-free) definition that can distinguish "cause" from "effect" without simply referring to the order in which events happen. Just because an idea is intuitive does not mean that it can be properly translated into a physical model.

Even after selecting for (mathematical) self-consistency, our intuition plays a role in determining which ideas we try out first. As we know, only a smattering of physical models seem to match reality. If our intuition is guiding us in the wrong direction (as the above example certainly suggests), this will slow our exploration of the ideas that will actually pass experimental muster. The fact that experimental confirmation has given an (always-provisional) stamp of approval to many counter-intuitive physical theories is more evidence that our intuitions have guided us astray.

Given our mistaken intuitions, then, it should be no surprise that we're also mistaken about which types of mathematics we might expect to be the most useful for fundamental physics. Our intuition may get a vote, but it's the universe that has rigged the election, and is apparently following (self-consistent) mathematical rules that seem strange and esoteric to us humans. Outside of our most fundamental physics, different mathematical rules can be useful in describing new higher-level phenomena—and there's no guarantee that our most fundamental physics is truly fundamental at all.

The last mystery addressed here is why mathematical pursuits often discover the outlines of strange mathematical structures before they have much hint of an application. Part of the answer is that we have more misleading physical intuitions

---

[5]It seems doubtful that we could develop a self-consistent set of mathematical axioms in which entities can meaningfully vary with respect to themselves.

than mathematical instincts. And here the *differences* between physics and math are highly relevant: mathematicians can be fearless explorers without being viewed as heterodox; physicists tend to wait for experimental reasons to venture in non-intuitive directions.

So while physics and math do have a striking degree of overlap, this is hardly some cosmic coincidence. The necessity for consistency in physical models, along with some mistaken human intuitions, can mostly explain the largest questions. It remains to be seen where physics goes next, but it seems likely that our models will find uses for even more unusual mathematical structures. The future of physics may lie in a counter-intuitive direction, but at least we know it will be framed in the language of mathematics.

Addendum

Here I summarize the most interesting discussion points that ensued after this essay was originally posted, on the FQXi forum.

One general point that was made several times in the comments (by Armin Shirazi and Jose Koshy, in particular), is that not all intuitions are misleading and harmful. Indeed, the carefully-honed intuitions of mathematicians and physicists are not only useful, but practically indispensible for making progress in both of these fields. This point is absolutely correct; my aim here was primarily the *innate* intuitions that humans have developed through evolutionary pressures. It is these, built-in-intuitions that have not been of particular benefit to fundamental physics for over a century, arguably doing more harm than good. So when reading the above essay, it is important to keep this qualification in mind when reading my use of the word "intuition".

That said, I would claim that some of the higher-level intuitions that we have developed over the years (in response to our successful theories) may still be misleading us in some ways. Unfortunately, the question of *which* of these intuitions are misleading is usually only clear in hindsight.

The remainder of this section consists of selected quotations from the comment sections (including other relevant essays, as noted). The most interesting discussion was with Matt Leifer, a second-prize winner in this contest.

Matt Leifer: Nice essay, and I am pleased to see that you have so elegantly managed to weave in another argument for the block universe. One thing that you have not addressed in your essay is the demarcation criteria for mathematics. Sure, mathematics proceeds from systems of axioms, but not all systems of axioms count as mathematics. There are many systems of axioms that would lead to boring or overly complicated theories, and mathematicians have a knack for avoiding exploring those ones.

ML: If mathematics just consisted of exploring all and any axiom systems then I agree with you that it would be no surprise to find that mathematicians had already explored the structures needed for physics. However, in actual fact, mathematicians have a lot of guiding principles for what constitutes a mathematical theory, including elegance, unifying power etc. One has to explain why the mathematics developed according to those principles is likely to show up in physics, rather than just any axiom systems. My answer to this is that mathematicians' criteria for what counts as mathematics ultimately stem from the natural world, so it should be no surprise that those theories later show up in our descriptions of the natural world. I would like to know what you think about this.

KW: Interesting point... I'm far more willing to accept your point on the "overly complicated" axioms than the "boring" ones. I would imagine that the reason certain axioms lead to 'boring' results is that they're isomorphic to some other well-understood bit of mathematics.

The following quotations, on this page, come from my subsequent response to Matt Leifer's essay [5], *and responses on that thread.*

KW: I think you're absolutely right that some new mathematical "nodes" are developed for exactly the reasons you describe here. But is this the *only* way that new mathematical "nodes" are developed? It seems to me that an equally-important process is simple exploration/extrapolation from existing nodes, tweaking the premises that lead to one bit of math, and seeing what new bits might result. Sure, this isn't going to necessarily lead to the efficient network structure that you describe, but it would supply a bunch of "raw material" which your efficiency-driven process could then organize.

KW: One big distinction between the nodes generated by an "exploration process" (rather than your "efficiency process") is that there's generally no requirement that they hold together self-consistently. This is what I had in mind when I talked about different "caves" in my own essay, even though you're right that eventually your efficiency-driven process tends to merge the caves together into much larger networks... maybe even to the point where you can claim that there is one "pragmatically-true" type of math when it comes to things like the axiom of choice.

KW: I think your comment on my essay reflects that we're trying to answer two somewhat different interesting questions. You are addressing the mystery of why the most-efficient network groups of math tend to map to physics; I'm trying to address the mystery of why so many separate explorative-nodes tend to find a use in physics.

ML: I do not think that much "pure" exploring actually goes on in mathematics. If that's what mathematicians were really doing then why would they not just explore arbitrary axiom systems? Mathematicians typically identify two strands of mathematical work: theory-building and problem solving. In reality, these are not completely separate from each other. A theory may need to be built in the service of solving a problem and vice versa. As an example of the former, consider the $P = NP$ problem. Someone notices that although they cannot prove a separation directly, they can if they introduce a modified model of computing with some funny class of oracles. After that, they start exploring this new model, partly because it is expected to eventually tell on the $P = NP$ problem, but also perhaps because it has its own internal elegance and might have practical applications different from the problem for which it was originally invented to solve. I think the vast majority of exploration/extrapolation is of this type. It does not exist in a vacuum, but is closely tied to the existing structure of knowledge and the existing goals of mathematicians.

ML: You are right that there are other processes going on in the knowledge network other than the replacement of analogies with new hubs. The latter would be a process of pure theory building, whereas in reality there is a mix of theory building and problem solving going on. I emphasized the theory-building process primarily because I think it is key to explaining why abstract mathematics shows up in physics.

KW: Perhaps the only real daylight between us might then be addressed by the following thought experiment. Suppose that some group of mathematicians decided to do "pure exploring", to no problem-solving purpose, and then came up with some new fields of mathematics that weren't directly connected to anything that had come before. (Although of course your efficiency process might later find deeper analogies with known mathematics.) Is it your contention that such theories would be much less likely to find use in future physics, as compared to theories generated by a problem-solving-style motivation?

KW: I'd be surprised if this were true. So long as the "pure-exploration" theories were self-consistent and of comparable complexity, I'd think they would be just as likely to find use in physics as theories that had been inspired by some purposeful, problem-solving process. So for me, it doesn't particularly matter whether the "generalization" process loses the root-level link to empirical reality; that link can be re-established at a higher level, if physics finds a given theory useful for some real-world problem.

Alexey Burov: Pondering on your conclusion, I asked myself: writing this, did you keep in mind that "the laws of nature are described by beautiful equations", as Wigner's brother-in-law put it? If yes, how might this explanation look like? If not, wouldn't the key part in the physics-mathematics relation be lost?

KW: I agree with your sentiment, but unfortunately "beauty" is a bit too subjective of a premise to start with when looking for an objective explanation. Even "simplicity" and "elegance" have subjective aspects, but maybe "efficient" is the right starting point. We humans (or at least some of us) find it beautiful when a very wide range of phenomena can be explained with a few efficient concepts. Asking why this is in fact the case is an excellent question, but I wouldn't say it's the key part of the mystery.

KW: I say this because even if I provided a good explanation for why there are a few rules that explain everything, that would really only apply to the most fundamental physics from which everything else emerges. Such an explanation wouldn't cover higher-level, effective- or emergent-physics, for which mathematics is certainly still important, and this mysterious overlap between math and physics continues. I would say that the most use of higher-level math actually takes place at this higher level, where any "ultimate efficiency" arguments don't really apply.

KW: Furthermore, I'm convinced we haven't gotten down to the truly fundamental level yet, in any of our theories except maybe perhaps general relativity. So at this point, I see pretty much all of physics as a higher-level approximation, and speculating about the efficiencies of the fundamental level that may be waiting for a discovery is just... well... speculation! Although I am convinced, as are most physicists, that any ultimate explanation will indeed be efficient.

Laurence Hitterdale: The standard metaphors for the distinctive feature or features of time are in some ways more misleading than illuminating. I can agree with your critique of the images of flow and passage. Nonetheless, I would contend that the lack of a mathematical representation should not be taken as leading to the conclusion that the supposed distinctive characteristics of time are non-existent. The warranted conclusion is, I think, somewhat more complicated: either those characteristics do not exist, or they exist in the physical world without being mathematically representable, or they exist only subjectively, that is, only in experience.

LH: On the last alternative, it remains true that the distinctive features of time are real, even though they are not part of physical reality. In that case, any attempt at a complete account of the nature of things would still be under the obligation of trying to explain those features... Time is an experienced part of reality, and it is experienced as something very different from a dimension of space. I do not know how the experience arises. Maybe it derives from other entities and forces which in themselves lack distinctive temporality.

KW: You give 3 options; the first and last I'm okay with. But not "they exist in the physical world without being mathematically representable". If there's no consistent mathematical framework in which they can be discussed, then the very notion is inherently inconsistent and won't have a physical counterpart.

KW: As for the last option, that our perception of time is real as far as *experience* is concerned, that's fine, but that makes this issue a consciousness-problem, not a physics problem. I wish physicists would recognize this and leave it alone. (Unless, I suppose, they're also going to be building useable models of consciousness... but that's still not physics.) Looking to "other entities and forces" to explain one aspect of our conscious experience seems to be like a huge mistake, mixing lower-level and higher-level concepts in a way that seems wholly and utterly implausible—especially because those "other entities" don't seem to show up at any of the intermediate levels between physics and consciousness (mesoscopic physics, chemistry, biology, neural networks, etc.).

Peter Jackson: You write; "The only caves that physicists find worth entering are the structurally-sound ones that have been mapped out by the mathematicians." This infers that that no worthwhile hypothesis can be conjectured unless the maths preexist it. More worrying it seems to imply there's no room for a physical mechanistic hypothesis to be developed ('intuitive' or not) which can then be used to derive the mathematical description.

KW: I certainly didn't mean to imply this; I do think it's generally true, but not necessarily true. Hopefully I made it clear eventually that one might develop new mathematics to handle new hypotheses. So no, it's not an *absolute* pre-requisite.

PJ: You then again seem to leave the solid ground and suggest it's more likely; "math is just less biased than physics, more willing to explore new ideas." Which I must admit is counter to most of my experience. Do you suggest that's my intuition being wrong?

KW: It's not counter to mine, but then again I might be overly biased, as I've been finding it hard to push unusual physics ideas. I would say that when it comes to dramatic changes, leaping into some quite different framework, math is more fearless than physics, for the reasons I outline in the essay.

PJ: Nowhere do you identify that apparently self consistent maths can contain flaws and mislead us or not fully correspond to the physical processes for which it's invoked.

KW: I guess I didn't dwell on the fact that most mathematically-consistent physics ideas don't pan out, but that's absolutely true. (It's not the math that misleads us; it's the incorrect physics!)

PJ: I also agree time doesn't flow and has no 'direction'. It seems a horrendous 'category error' to endow it with material attributes. But does that example prove that intellectual rationalisation, deduction and induction (intuition?) are near valueless any more than fallacious mathematical proofs show that maths is useless?

KW: Just because our innate intuitions are near-valueless when it comes to deep theoretical physics doesn't mean that we can't build up new intuitions that *are* useful. What the example shows is that math can steer us away from making serious mistakes, even if our innate intuitions insist on making them. (In the same way, math can tell us which are the fallacious proofs, even if several people make the same mistake.)

PJ: Of course I agree your basis premise that maths is an (one) essential tool and 'consistency check' but is it the only one?

KW: Certainly not! More important by far is experimental verification.

PJ: As Alma points out in her excellent essay, in mathematics any inconsistency would be shot dead on sight. In physics this often isn't the case and inconsistencies become acceptable by familiarity! I think that may be the greatest danger of the assumptions you hypothesize, perhaps responsible for ever deeper theoretical entrenchment.

KW: Very interesting point... And I worry greatly about people accepting inconsistencies due to familiarity, especially in quantum theory. Maybe physics should strive to be a lot more like math in this regard, along with being more willing to explore a wider parameter space of ideas.

References

1. E. P. Wigner, Comm. Pure App. Math. **13** 1–14 (1960).
2. A.C. de la Torre and R. Zamora, R., Am. J. Phys. **69** 103 (2001).
3. S. Pinker, "The Stuff of Thought", Penguin (New York), chapter 4, (2007) and references therein.
4. W. Craig and S. Weinstein, Proc. Royal Soc. A **465** 3023–3046 (2009) and references therein.
5. M. S. Leifer, "Mathematics is Physics", http://fqxi.org/community/forum/topic/2364, and also in this volume.

How Not to Factor a Miracle

Derek K. Wise

Abstract Wigner's famous and influential claim that mathematics is "unreasonably effective" in physics is founded on unreasonable assumptions about the nature of mathematics and its independence of physics. Here I argue that what is surprising is not the effectiveness of mathematics but the amenability of physics to reductionist strategies. I also argue that while our luck may run out on the effectiveness of reductionism, mathematics is still our best hope for surpassing this obstacle. While I agree that human understanding of the natural world in mathematical terms evinces a miracle, I see no way to factor out the human dimension of this miracle.

Mathematics is a bit like Zen, in that its greatest masters are likely to deny there being any succinct expression of what it is. It may seem ironic that the one subject which demands absolute precision in its definitions would itself defy definition, but the truth is, we are still figuring out what mathematics is. And the only real way to figure this out is to *do* mathematics.

Mastering any subject takes years of dedication, but mathematics takes this a step further: it takes years before one even begins to see what it is that one has spent so long mastering. I say "begins to see" because so far I have no reason to suspect this process terminates. Neither do wiser and more experienced mathematicians I have talked to.

In this spirit, for example, *The Princeton Companion to Mathematics* [6], expressly renounces any tidy answer to the question "What is mathematics?" Instead, the book replies to this question with 1000 pages of expositions of topics *within* mathematics, all written by top experts in their own subfields. This is a wise approach: a shorter answer would be not just incomplete, but necessarily misleading.

D.K. Wise (✉)
present address
Concordia University,
Saint Paul 1282 Concordia Ave, 55104 St. Paul, Minnesota, USA
derek.k.wise@gmail.com

Department Mathematik, Universität Erlangen-Nürnberg,
Cauerstraße 11, D-91058 Erlangen, Germany

A. Aguirre et al. (eds.), *Trick or Truth?*, The Frontiers Collection,
DOI 10.1007/978-3-319-27495-9_10

Unfortunately, while mathematicians are often reluctant to define mathematics, others are not. In 1960, despite having made his own mathematically significant contributions, physicist Eugene Wigner defined mathematics as "the science of skillful operations with concepts and rules invented just for this purpose" [8]. This rather negative characterization of mathematics may have been partly tongue-in-cheek, but he took it seriously enough to build upon it an argument that mathematics is "unreasonably effective" in the natural sciences—an argument which has been unreasonably influential among scientists ever since.

What weight we attach to Wigner's claim, and the view of mathematics it promotes, has both metaphysical and practical implications for the progress of mathematics and physics. If the effectiveness of mathematics in physics is a 'miracle,' then this miracle may well run out. In this case, we are justified in keeping the two subjects 'separate' and hoping our luck continues. If, on the other hand, they are deeply and rationally related, then this surely has consequences for how we should do research at the interface.

In fact, I shall argue that what has so far been unreasonably effective is not mathematics but *reductionism*—the practice of inferring behavior of a complex problem by isolating and solving manageable 'subproblems'—and that physics may be reaching the limits of effectiveness of the reductionist approach. In this case, *mathematics* will remain our best hope for progress in physics, by finding precise ways to go beyond reductionist tactics.

Is Physics Unreasonably Well Described in the Language Invented for that Purpose?

The essence of Wigner's claim, based an unsupported characterization of mathematics as an elaborate mental game that has nothing a priori to do with physics, is that mathematics is somehow nonetheless amazingly effective in describing physics. Were physics and mathematics disjoint endeavors, one empirical, one cognitive and largely arbitrary, then it would indeed be astounding, even 'miraculous,' to find them interacting as fruitfully and precisely as they do. But this viewpoint disregards both history and the mathematical biases set up by our place in the natural world.

First, historically, the distinction between mathematics and physics is relatively recent, with the movement toward 'pure' mathematics progressing gradually since the 18th century. The historical counterargument to Wigner's thesis has been taken up by mathematics historians [7] in detail, and that is not my purpose here. However, it is worth pointing out that, for example, calculus and a large part of differential equations—still among the physicist's most important tools—were designed precisely for physical applications. It is at least anachronistic, but bordering on absurd, to claim these are unduly effective in the very subject they owe their existence to.

In fact, the joint development of physics and mathematics predates the recognition of physics as a separate science. I must agree with Einstein that "mathematics generally, and particularly geometry, owes its existence to the need which was felt of learning something about the relations of real things to one another," and that "we may in fact regard [geometry] as the most ancient branch of physics" [3].

Second, at a deep level, mathematics—or rather our limited view of it—is founded on our perception of the natural world. To get an idea of how much our view of mathematics depends on physics, we need not imagine a world with drastically different physical laws: we can simply consider our own world at different scales, where subtle physical phenomena become readily apparent.

If we were very *large* mathematicians, or equivalently if Newton's gravitational constant G were large, Einstein's famous curving of space and time would be noticeable on the scale of everyday life. With no rigid spatial backdrop, and no independence of 'space' and 'time' on ordinary scales, we would at least have a very different history of geometry.

Mathematics would be much more radically different if instead we were very *small* mathematicians, or equivalently if Planck's constant $\hbar$, which regulates the level of 'quantum fuzziness,' were large. This would also have affected our ideas of geometry, since there would be no stationary objects with well-defined positions to serve as reference points for measurements. But also, we may well have developed little interest in numbers or counting, since distinct collections of objects become much less sensible at quantum scales. We would thus have less motivation to invent number theory, set theory, or combinatorics. These subjects, fundamental as they are to mathematics as we know it, enjoy this status because of our place within the natural world.

Moreover, as 'quantum mathematicians' we would presumably prove theorems not according to the familiar rules of logic, which are deeply tied to set theory, but according to rules of quantum logic, which reflect the fuzzy, indistinct nature of propositions about fundamentally quantum mechanical systems. Likewise, our foundational notions of probability and statistics would necessarily be quantum in nature.

More radical still, we can consider a scenario in which G and $\hbar$ are both large, corresponding to a realm of experience where effects of 'quantum gravity' should be evident. Here the foundations of mathematics would surely differ even more radically from our usual ones, in ways that one can only speculate on depending on which approach to quantum gravity, if any, one believes.

While it is interesting to consider how our foundations of mathematics are determined by physics, it is in fact recognized *from a perspective entirely within pure mathematics* that mathematics itself could be built on alternative foundations. This goes beyond the philosophy of mathematics to what we might call the *mathematics of mathematics*, in that the types of foundations of mathematics—different 'mathematical worlds' with their own notions of logic—can be classified mathematically. For example, the potential 'mathematical foundations' with many of the same formal properties as the usual set-based foundations are classified by topos theory. This motivates attempts to use topos theory in the foundations of physics [5].

If our experience of the physical world were different, then we would have invented very different mathematics, but it would still be mathematics. It would still be quite effective at describing the world it was built to represent—presumably even more effective in treating certain phenomena.

Appropriate Abstraction

Of course, one might object that while some mathematics has been invented specifically for physical applications, it remains surprising that 'abstract' mathematics seems to keep finding its way back into physics. In fact, this confusion arises when we erroneously conflate the precise definition of 'abstract' with its colloquial one.

Outside of mathematical or philosophical discussions, when people call a thing 'abstract,' they may only mean it is difficult to understand, or even unnecessarily cerebral. But true abstraction is just the opposite: a path to *clarity*. It means stripping away all inessential details, arriving at an idealization whose life purpose is to exemplify the properties or concepts under consideration, and nothing else. Russell used the word in this sense when he said the language of physics is necessarily abstract, in order to "say as little as the physicist means to say."

This raises a key question: How does one arrive at a *good* mathematical abstraction? It is perhaps not surprising that we can find useful mathematical abstractions of the *simplest* objects in our experience; ultimately, our ability to find good mathematical abstractions of concepts in physics, especially as compared to other sciences, must rest on physics being much simpler than other sciences.

Much more surprising—and I gather this is the part Wigner also finds incredible—is the ability to 'stay' at an abstract mathematical level, not referring back to direct physical experience, and still somehow manage to arrive at new mathematical constructs which later turn out to have applications to physics.

So, let us ask: Where do new mathematical concepts come from? Sadly, Wigner is again rather dismissive on this point, claiming that most advanced mathematical concepts are defined just so that the mathematician can "demonstrate his ingenuity and sense of formal beauty [8]."

In fact, defining new concepts in mathematics is a delicate art, and evidently one of the hardest to master. Whereas a decent Ph.D. student in mathematics will be able to prove some quite difficult theorems—this is the most obvious skill that a graduate education in mathematics teaches—there is typically no training on what makes a good definition. Finding the right definitions is simply too advanced. Some definitions, once one recognizes a need for them, are rather obvious. But in many cases, it takes years for a community of mathematicians to agree on the best definition for a concept.

To understand what makes a good mathematical definition, it is of course best to study many examples. Here I will consider just one, but deliberately choose an example that does not yet have well-established physical applications.

One of my current favorite examples is the notion of *quantum groupoid*, also called a Hopf algebroid [1, 2]. Groupoids are a modern mathematical way to study *symmetry*—arguably the most important concept in mathematical physics. Intuitively, quantum groupoids result from taking the concept of groupoid, based on sets and hence classical logic, and importing it into the world of vector spaces and hence quantum logic.

But realizing this intuitive idea and settling on a definition of quantum groupoid took hard work. The definition that finally emerged looks quite complicated and, frankly, esoteric from the perspective of an outsider showing up on the scene after the dust has settled. It is much more complicated than what I would instinctively try, or what researchers did in fact try first. If you handed me the accepted definition of quantum groupoid out of the blue, I would think it very unlikely to be useful in physics.

So, what makes this definition 'good' or 'correct' in some sense? Consider this evidence:

- The definition naturally relates mathematical structures with many known applications, including 'groupoids' and 'quantum groups.'
- More naive attempts at the definition were refined by the need to include natural examples.
- Despite being rather complicated, equivalent definitions were discovered *independently* by researchers with quite different starting points, from within different subfields of mathematics.

Notice that all of this evidence for the definition's quality (and more could be listed) is purely mathematical, and none of it is based on our desire to "demonstrate [our] ingenuity and sense of formal beauty." On the contrary, the evidence gives a strong sense that this is a robust mathematical concept. Together with the tight relationship to concepts with known physical applications which one may naturally want to combine, this makes me think physical applications for quantum groupoids may not be so far-fetched after all.

Ways of Ignoring

Of course, the simpler a physical system is, the more likely we can find an appropriate mathematical abstraction of it, and in practice we do not model the entire system of interest but some simplification of it. The idea that we can study a complicated system by isolating certain aspects of it to focus on is known as *reductionism*. This brings us to what I argue has truly been 'unreasonaly effective' in physics.

We will need to distinguish between two types of reductionism. At a basic level, reductionism means *ignoring* nearly everything in the universe. But there are different ways of ignoring: it is one thing to ignore most of the *things* in the universe, and quite another to ignore most of the *properties* of the things in the universe. For clarity, I will call the first of these **reductionism** and the second **co-reductionism**. Both play foundational roles in physics, so it is worth taking some time to understand them in detail.

Mathematically, the difference between these two kinds of ignoring corresponds to the difference between *sub*-objects and *quotient* objects. A sub-object is essentially what it sounds like: a smaller piece of some other object. A particular *sub*set of set of all people is the set containing only Eugene Wigner, and this can be represented as an arrow from the subset into the whole set, indicating how the part is included in the whole:

We engage in *reductionism* when we focus in on a sub-object, for example, to study Wigner as an individual. Of course, if our goal is to write a biography, we care not just about the Wigner but also his place in the set of all people—that is where the arrow comes in. However, our main subject is Wigner, and we ignore all other people at least insofar as they do not interact with Wigner.

Quotient objects are less familiar, but are easy to understand by examples. A particular *quotient* of the set of all people is the set of genders. This can be represented as arrow going from the whole set to the quotient set:

The arrow simply represents assignment of the appropriate gender to each person. Here we are not ignoring *any* people, but rather ignoring nearly everything *about* all of them.

Mathematicians have a nice way to think of this: the arrow is thought of as a process in which we "identify"—declare to be identical—any two source elements that map to the same target element. In the example, we imagine that Wigner and I, by virtue of being both male, are identified: we *become the same person* according to a vastly oversimplified model where the only distinguishing feature of a person is gender.

It is important to distinguish between 'sub-' and 'quotient' constructions. For example, ♀ and ♂ do not refer to particular people, so it would be wrong to think of {♀, ♂} as a 'subset' of the set of people; it is a quotient. (Confusing this quotient with a subset may be the closest I can come to a mathematically rigorous definition of sexism.)

Reduction and Co-reduction in Fundamental Physics

Every field of study uses reductionism and co-reductionism, usually a mixture of both, for obvious reasons: It would be impossible to get off ground in any study without confining our attention to only some of the relevant things, or to some of their relevant properties.

Often, of necessity, we ignore what cannot be justifiably ignored, and settle for idealized models drastically simpler than the systems we are really interested in. While this remains a useful tool in all sciences, in physics it appears to be much more. If physicists are envied by other scientists, it is because the information they ignore has so often turned out to be perfectly justified, at least to within the limits of very precise measurements.

Reductionism in fudamental physics has succeeded because matter *really does* appear to be composed of indivisible and completely identical parts. While the properties of one person imply very little about people in general, the properties of one electron appear to tell us *everything* about all other electrons—the are indistinguishable in minutest detail. Beginning with the atomic theory, and continuing with the discovery of subatomic particles, and finally to the list of about 17 observed particles that physicists now take to be fundamental, we now have built up an astonishingly precise picture of matter—a reductionist's dream, with a finite set of building blocks and neat rules on how to combine them.

At the other end of the spectrum, what I have called *co*-reductionism plays a key role in the other fundamental pillar of modern physics, *general relativity*. In general relativity we ignore almost all properties of matter, considering only its mass. Remarkably, this again seems to be more than a practical convenience: gravity really does not seem to care about other properties of matter in the least.

What seems quite miraculous about this whole situation is that gravity, the one force that is so weak that one only notices it when there is a *lot* of matter around, is also the one force entirely unconcerned with any distinguishing properties of that matter: it seems to care only about mass and location of that mass.

Quantum gravity, the attempt to combine general relativity with quantum physics, faces a dilemma, since these theories involve orthogonal types of 'ignoring.' Most approaches to the problem can be classified into one of two scenarios. In one scenario, we try to make gravity more like the rest of physics, for example by studying elementary particles that carry the gravitational interaction ('gravitons') or frameworks to generalize them. In the other scenario, we attempt to apply principles of quantum physics, which in fact we know entirely from the world of particles, to the relativistic picture of space and time.

In other words, either we attempt to augment a theory arrived at by reductionist tactics to include a theory arrived at by co-reductionist tactics, or the other way around. Neither seems obviously bound to work. If our luck does run out, it will be a failure of reductionism, and not of mathematics. In fact, the precise and structural ways of thinking provided by mathematics will be our best hope to surpass this obstacle.

Factoring a Miracle

I have argued that what is truly surprising in physics is not that mathematics is effective in physics, but rather that reductionist strategies work so well, for apparently purely physical reasons. There is also a deeper sense of reductionism lying at the heart of Wigner's argument, and the worldview we are led to if we take it seriously.

On the surface, Wigner's claim that mathematics is unreasonably effective in physics seems related to Einstein's famous statement about the world being amazingly comprehensible. But taken in context, it is clear Einstein does not wonder at the comprehensibility of the world in some metaphysical sense, but rather that the world is comprehensible *to us*:

> That the totality of sensory experience is such that it can be organized through thinking ...is a fact that we can only marvel at, but which we will never be able to comprehend. We can say: the eternally incomprehensible thing about the world is its comprehensibility.

> Dass die Gesamtheit der Sinneserlebnisse so beschaffen ist, dass sie durch das Denken ...geordnet werden können, ist eine Tatsache, über die wir nur staunen, die wir aber niemals werden begreifen können. Man kann sagen: Das ewig Unbegreifliche an der Welt ist ihre Begreiflichkeit. [4]

While it is impressive that one can get so far *in principle* with extracting mathematical structure from the physical world, the shocking thing is that we have the mental capacity to do this *in practice*. The real miracle is the complexity of the world *relative* to our own intelligence. This is indeed cause to marvel. It is one thing for the universe to be sensible in some precise way, and quite another for some entity within the universe to *make* sense of it to the extent we have.

The desire to 'factor out' the human element no doubt stems from reductionist tendencies, where in this case the component we attempt to ignore is *ourselves*. I see no sensible way to do this.

Acknowledgments I am indebted to John Baez, Julian Barbour, David Corfield, and Paul Morris for conversations and insights that have clarified my thinking on this subject, perhaps especially in ways that did not impact this essay itself as much as they might have.

References

1. G. Böhm, Hopf Algebroids, in Handbook of Algebra Vol 6, ed. M. Hazewinkel, Elsevier, 2009, pp. 173–236.
2. B. Day, R. Street, Monoidal bicategories and Hopf algebroids, Adv. Math. 129 (1997) 99–157.
3. Albert Einstein, Geometry and Experience, Address to the Prussian Academy of Sciences, Berlin, Jan. 1921. English translation at http://www.history.mcs.st-and.ac.uk/Extras/Einstein_geometry.html.
4. Albert Einstein, Physik und Realität, J. Franklin Inst. 221 (1936) 313–347.
5. Chris J. Isham, Topos methods in the foundations of physics, in Deep Beauty: Understanding the Quantum World through Mathematical Innovation, ed. H. Halvorson, Cambridge University Press, Cambridge, 2011.
6. The Princeton Companion to Mathematics, eds. Timothy Gowers, June Barrow-Green, Imre Leader, Princeton University Press, New Jersey, 2008.

7. Ivor Grattan-Guinness, Solving Wigner's mystery: The reasonable (though perhaps limited) effectiveness of mathematics in the natural sciences, The Mathematical Intelligencer 30 (2008) no. 3, 7–17.
8. Eugene Wigner, The unreasonable effectiveness of mathematics in the natural sciences, Comm. Pure Appl. Math. 13 (1960) 1–14.

The Language of Nature

David Garfinkle

In 1960 Eugene Wigner published an article titled "The Unreasonable Effectiveness of Mathematics in the Natural Sciences" [1]. He gave several examples of areas of mathematics that had developed independently of physics, but that nonetheless proved to be essential in the formulation of twentieth century physics. Wigner concluded that "The miracle of the appropriateness of the language of mathematics for the formulation of the laws of physics is a wonderful gift which we neither understand nor deserve."

Wigner's view stands in contrast to the view of Galileo in which he used the image of nature and its laws as a book and asserted that that book was written in the language of mathematics [2]. In Galileo's view, when we develop physical theories we are discovering the underlying mathematical order of nature. I will argue that Galileo's view is essentially correct, but that the developments that led to Wigner's view reveal two deep truths: one about the nature of mathematics and the other about the nature of physics.

To begin, it is helpful to recall the development of non-Euclidean geometry [3]. Though expressed in an axiomatic form by Euclid, geometry can be thought of as something akin to a physical theory, with the properties of its lines revealing empirically discovered properties of stretched strings, or the lines of sight of surveying and thus the paths of light rays. However, of the axioms used by Euclid, one seemed less natural than the others: the parallel postulate, which is the statement that given a line and a point not on the line there is exactly one line through the given point that is parallel to the given line (Euclid did not put the parallel postulate in this form; but what he used is equivalent to this). Because the parallel postulate seemed unnatural, there were efforts to derive it from the other axioms of Euclidean geometry, but all such efforts were unsuccessful.

Finally in the 19th century Gauss, Bolyai, and Lobachevsky went in the opposite direction, producing a geometry that used the other axioms of Euclid, but in which

D. Garfinkle (✉)
Department of Physics, Oakland University, Rochester, MI 48309, USA
e-mail: garfinkl@oakland.edu

A. Aguirre et al. (eds.), *Trick or Truth?*, The Frontiers Collection,
DOI 10.1007/978-3-319-27495-9_11

the parallel postulate was replaced by the statement that through the given point there was more than one line parallel to the given line. In modern language, we would say that Gauss, Bolyai, and Lobachevsky had developed the geometry of the hyperbolic plane: a two dimensional space of constant negative curvature.

The development of non-Euclidean geometry marked the divergence of mathematics from physics. Euclidean geometry and non-Euclidean geometry gave different answers from each other, but they were both "right" in the sense that both were consistent systems derived by logic from a set of axioms. Though one could argue about which geometry better modeled the physical world, as long as one uses only consistency as the criterion, the two geometries were on an equal footing. No longer was mathematics to be about finding the "right" axioms (i.e. the ones that seem to come from nature). Instead the mathematician could choose any set of axioms he (or she) liked as long as they were consistent, and then whatever followed from those axioms would be mathematics.

Partly in response to this new found insight, the study of mathematics exploded: more and more systems of greater and greater abstraction and complexity were developed, and the features of these systems were worked out in meticulous detail. This trend continued through the 20th century, with mathematics becoming ever more formal and abstract, and the number of different systems studied by mathematicians ever increasing.

The 20th century also saw the development of new and revolutionary theories of physics: relativity and quantum mechanics. And, miraculously, when the new theories of physics needed new mathematics, lo and behold that new mathematics had *already* been developed: differential geometry for the general theory of relativity, group theory, complex vector spaces and their operators for quantum mechanics. Furthermore, this extraordinarily useful mathematics had been developed by mathematicians who had merely been pursuing abstract subjects with total disregard for any physical meaning or application those subjects might have.

It is that extraordinary coincidence, that the new physics needed new mathematics, *and there it was*, that seemed so miraculous to Wigner. But I will now argue that this development is not as miraculous as it seems. First, let us consider the nature of the new mathematics. Non-Euclidean geometry and the formal and axiomatic approach to mathematics gave the mathematicians enormous freedom. Any formal system of axioms could be considered mathematics: simply develop a system of axioms (as long as they are consistent with each other) and then these axioms along with all theorems derived from them form a mathematical system. But what were the mathematicians to do with all of this newfound freedom? If anything can be mathematics, then how is a mathematician to choose what to work on?

Wigner claimed that mathematicians chose systems and problems solely to exhibit their cleverness in making formal arguments. Well, perhaps that claim has some element of truth to it, and it is not surprising that Wigner, who was a close friend of John von Neumann since childhood, would have that view of mathematics and mathematicians. But just as "choose any set of axioms you like" does not give sufficient guidance in the development of new mathematics, so "exhibit your mathematical cleverness" is also too amorphous a criterion to be a guide in the development of new mathematics.

So how did the mathematicians decide what would be the new mathematics? They did it by a process of abstraction and generalization from the old mathematics: group theory to generalize the properties of the symmetries of objects in Euclidean geometry; differential geometry to generalize curved surfaces in three dimensional Euclidean space; operators and vector spaces to generalize both the properties of the usual vectors of three dimensional Euclidean space and the properties of linear differential equations and their solutions; analysis and topology to abstract and generalize a set of useful techniques for proving the convergence of series, especially those used to finally put calculus on a rigorous footing. This process of abstraction and generalization generated a great number of new mathematical objects and led to another pursuit of modern mathematicians: classification. For each new type of object (group, vector space, manifold, Lie Algebra, etc.) one would aim to produce a complete classification of all possible objects of that type. The process of abstraction and generalization can be thought of as exploring the realm of possible new mathematical systems by starting at the old systems and working one's way outward. The process of complete classification means that in a certain sense this exploration process is very thorough: that any new mathematics that is "sufficiently close" to the old mathematics will be found.

Thus the new mathematics was related to the old mathematics, which was in turn related to the old physics. But why was the new mathematics just what was needed for the new physics? Here the answer has to do with the fact that old physical theories are limiting cases of new physical theories. To cite some well-known examples: Newtonian mechanics is the limit of special relativistic mechanics in the case of small velocity; special relativity is the limit of general relativity in the case of weak gravity (and thus Newtonian gravity is the limit of general relativity in the case of small velocity and weak gravity); classical mechanics is the limit of quantum mechanics for large systems (or more specifically actions large compared to Planck's constant).

But why do we discover the limiting cases first? Because of limited data and Ockham's razor. We, and the things we move and throw, are slow moving objects (compared to the speed of light) and so the most easily accessible data for us on the motion of objects is well described by Newtonian mechanics. But Ockham's razor enjoins us to prefer simple theories to complicated ones, and so Newtonian mechanics was developed before special relativity. We (and the things we move and throw) are large objects, compared to the size of atoms, and so classical mechanics is a good description of the data that was accessible to our experiments before the end of the 19th century, and therefore classical mechanics developed before quantum mechanics.

When new data becomes available, data no longer well described by the old theory, we must develop a new theory. But what new mathematics is needed for the new theory? If the old theory is a limiting case of the new theory, then it is likely that the mathematics of the new theory has something in common with the mathematics of the old theory. In particular, the new mathematics is likely to be some sort of generalization of the old mathematics. But generalization (along with abstraction) is precisely the business of contemporary mathematicians, as it was for the mathematicians of the 19th and 20th century ever since the development of non-Euclidean geometry. Furthermore, since a thorough classification of the possible newly discovered mathematical objects is also the business of mathematicians, it is likely that the particular generalization of the old mathematics needed by the new physics would be one that the mathematicians would discover, even before it was needed. Thus, we should not be surprised that when a new physical theory is developed, the new mathematics that is needed for that new theory is already at hand.

One of the things that struck Wigner as particularly miraculous is the central role of complex numbers in quantum mechanics. Complex numbers were discovered hundreds of years before quantum mechanics as a mathematical trick for finding roots of polynomial equations. But why should a trick for finding roots lie at the heart of our deepest theories of nature? It seems to me that the answer lies in the mathematical concept of a field. Roughly speaking, a field is a number system that has the additive and multiplicative properties of the real numbers. However, the mathematical classification of fields shows that the field concept is very restrictive: there are very few fields, even fewer that contain the real numbers, and of those fields that contain the real numbers the simplest one is the complex numbers. Thus one can think of the complex numbers as the minimal extension of the real numbers. It is therefore not so surprising that when studying the algebraic properties of real numbers, mathematicians stumbled across this minimal extension. And it is also not so surprising that for a system of nature based on the complex numbers, there are limiting cases that use only the real numbers and that physicists found those limiting cases first.

In future developments in theoretical physics, there is likely to be the need for some new mathematics. I will now argue that when that occurs, the effectiveness of the new mathematics will not seem miraculous. Here it seems to me that the situation is somewhat obscured by the present practice of string theory, in which much exotic mathematics is used and in which each new development is hailed as a breakthrough, if not a revolution. However, mathematics is effective in physics (let alone unreasonably effective) when it is an essential part of a theory that has been confirmed by experiment, not when it is part of the formalism of a highly speculative theory that has not been so confirmed. I predict that eventually the current mania for string theory will die down and that in its wake much of the mathematics that it uses will be regarded as not having a role in the description of nature. In the meantime, the mathematicians will go on to develop many new areas of mathematics, so that the subset of mathematics that is used effectively in physics will become an ever smaller fraction of the total amount of mathematics. At that point, it is unlikely to appear miraculous that the particular tiny subset of mathematics used in the description of nature is appropriate for that purpose. Thus it seems to me that the seemingly unreasonable effectiveness of mathematics was a one-time historical phenomenon of the twentieth century: it occurred shortly after physics and mathematics went their separate ways. Thus in that short time (1) mathematicians had developed some mathematics without regard for any physical application it might have, and (2) the new mathematics was developed by generalizing parts of mathematics that were closely related to physics, so that (3) when new physical theories needed new mathematics there was a good chance that this closely related mathematics would be just what was needed. As we get further and further from the 19th century parting of the ways of physics and mathematics, it is likely that these two fields will have less and less in common, and what little overlap there is will seem fortuitous rather than miraculous.

In summary the "unreasonable" effectiveness of mathematics is not so unreasonable after all. Nature and nature's laws are mathematical, just as Galileo and Newton taught us. It is our job as theoretical physicists to discover those laws. When new laws are discovered, they may use mathematics that has not been used in physics before. If so, we should not be surprised if that mathematics has already been invented by mathematicians for their own purposes. New physics has old physics as a limiting case. Thus it is not too surprising that the mathematics needed for the new physics is something that can be found by abstracting and generalizing the mathematics used in the old physics. Therefore the mathematicians, who develop new mathematics by abstracting and generalizing old mathematics, may develop just what we need even before we need it. This is indeed, as Wigner said, a wonderful gift; but perhaps we do understand it after all.

References

1. E. Wigner, Communications in Pure and Applied Mathematics, 13 (1960).
2. Galileo Galilei The Assayer.
3. For a short summary of the history and properties of hyperbolic geometry see e.g. Chapter 1 of S. Weinberg Gravitation and Cosmology, Principles and Applications of the General Theory of Relativity (1972) (John Wiley and Sons, New York).

Demystifying the Applicability of Mathematics

Nicolas Fillion

Abstract Essential tensions remain in our understanding of the reasons underlying the striking success achieved in science by applying mathematics. Wigner and many likeminded scientists and philosophers conclude that this success is a miracle, a "wonderful gift which we neither deserve nor understand." This essay seeks to dissipate that aura of mystery and bring the factors underlying the success of applied mathematics into the fold of scientific rationality.

Inquiries into the nature of mathematics as a science of its own and into its role in empirical science have a venerable tradition. Given that mathematics displays a kind of exactness and necessity that appears to be in sharp contrast with the contingent character of worldly facts, the problem that is perhaps the most unsettling examines how mathematics can be used to adequately represent the world. For instance, Einstein argued that "[t]he laws of mathematics, as far as they refer to reality, are not certain, and as far as they are certain, do not refer to reality" [6]. Similarly, Russell maintained that "[t]he exactness of mathematics is an abstract logical exactness which is lost as soon as mathematical reasoning is applied to the actual world" [8]. And yet, since the scientific revolution, efforts devoted to writing the book of the world in the language of mathematics have been resoundingly successful.

In light of this tension, many scientists and philosophers maintain that the applicability of mathematics is condemned to remain intrinsically mysterious. For instance, Wigner famously claimed that the "miracle of the appropriateness of the language of mathematics for the formulation of the laws of physics is a wonderful gift which we neither understand nor deserve" [10]. Dirac has similarly claimed that "[t]here is no logical reason why [the method of mathematical reasoning to study natural phenomena] should be possible at all, but one has found in practice that it does work

N. Fillion (✉)
Department of Philosophy, Simon Fraser University,
4604 Diamond Building, 8888 University Drive, Burnaby, BC V5A 1S6, Canada
e-mail: nfillion@sfu.ca
URL:http://www.nfillion.com

A. Aguirre et al. (eds.), *Trick or Truth?*, The Frontiers Collection,
DOI 10.1007/978-3-319-27495-9_12

and meets with remarkable success" [5]. It is true that, due to resilient tensions in our understanding, the applicability of mathematics is surrounded by an aura of mystery, but the present essay seeks to bring it back into the fold of scientific rationality.

A Mosaic of Problems

Failure to make significant progress towards solving a foundational problem often results from a clumsy understanding of that problem. In the case of the applicability of mathematics, it is also the case that part of the mystery stems from gathering many problems that require different types of solutions under the same umbrella. The striking achievement we wish to explain is the *success* of our use of mathematics in scientific practice. Yet many of the most widely discussed themes are only tenuously related to the explanandum.

Some such themes focus on mathematics qua language. There is bewilderment that it is even possible to use the language of mathematics to describe the world. In order to see that no mystery lies here, we must regard the activity of mathematical modelling as any other modelling practice. Constructing a model always involves the choice of a medium for the representation. Yet regardless of whether the medium chosen is plastic, wooden sticks, a picture, or statements in some language, models will succeed in capturing some aspects of a system, while other aspects will be idealized away. Each medium has its strengths and weaknesses. The main advantage of mathematics qua language is its considerable *expressive power* and *versatility*. If we consider the generality of foundational approaches to mathematics such as set theory or category theory, it would be difficult to imagine possible states of affairs that could not be somehow describable in mathematical terms. Hence the possibility of using the language of mathematics to describe the world is not in itself very surprising. But more importantly, the expressive power of mathematics qua language should not be conflated with our explanandum, for many mathematical expressions do *not* successfully apply. So what needs explaining are the circumstances that make *some* of the mathematics apply so successfully.

Another such theme focuses on the unexpected applicability of mathematical concepts developed in epistemic contexts in which no conceivable applications were anticipated. Yet such questions do not seek the actual reasons that underlie the successes of various applications. Instead they require an account of the conditions of possibility of such successes. However, in order to account for these actual successes, we would simply assume by fiat (based on the recent history of science) that mathematics *is* applicable, and then seek the causes of successful applications.

Thus, the problem is not one of characterizing mathematics as a language. Rather, it is one of explaining how to compare the virtues of different mathematical representations in a way that accounts for the success of those with a comparative advantage. It is common to identify truth as the theoretical virtue fulfilling this role. But even if we are willing to grant the broad point that the fundamental goal of science is the pursuit of truth, it cannot be denied that few of our successful models and theories are exactly true—indeed some are not even remotely true.

As I will argue in the next two sections, idealizations and other epistemic shortcuts play an indispensable role in the effective use of scientific rationality. For this reason, I take the main problem to be this: Given that the construction and manipulation of our successful mathematical models of reality is riddled with uncertainty, measurement error, modelling error, analytical approximations, computational approximations, and other forms of guesses and ignorance, how can their remarkable accuracy be explained? On the basis of the commonsensical "garbage in, garbage out" rule, this accuracy appears rather baffling, and accordingly I call this the problem of the *uncanny accuracy* of mathematics.

Too True to Be Good

As I have pointed out, an explanation of the actual success of applied mathematics in scientific practice is unlikely to be grounded in the literal truth of models, since very few models have this property de facto. However, to understand the nature of the success we seek to explain with more precision, it is important to acknowledge that this failure to be exactly true is a *feature*, not a bug. Indeed, any good theory idealizes away aspects of a physical system. Truesdell [9] elegantly make the point that "[o]ne good theory extracts and exaggerates some facets of the truth. Another good theory may idealize other facets. A theory cannot duplicate nature, for if it did so in all respects, it would be isomorphic to nature itself and hence useless." Theories and models play such a prominent role in physics because untangling the world is beyond the reach of our unmediated reason. We do not build theories to duplicate this complexity, but to set it aside as much as possible.

To illustrate this point, consider three different kinds of "idealized bodies" (or, idealized "building blocks") employed in classical mechanics to represent physical systems: mass-points particles, perfectly rigid bodies, and perfectly continuously deformable bodies. Despite their fundamentally idealized character, each type gives rise to a specific approach to classical mechanics. Articulating different idealized perspectives that complement each other enables us to efficiently get a grasp on the inner workings of physical systems. On the other hand, insisting on a *sub specie aeternitatis* true apprehension would be a path toward certain failure. A model or theory that contained "the whole truth and nothing but the truth" would quite simply be *too true to be good*.

Even so, if idealizations are to lead to any success, not any distortion can be warranted. Models should be *true enough* in order to be good. Mathematical modelling is a question-driven endeavour, so that the success has a pragmatic dimension. In applied mathematical practice, one considers real systems, i.e., systems as we actually encounter them in the universe we live in. So, in contrast to abstract models, a real model is not populated with mass-points, rigid bodies, or continuous media,

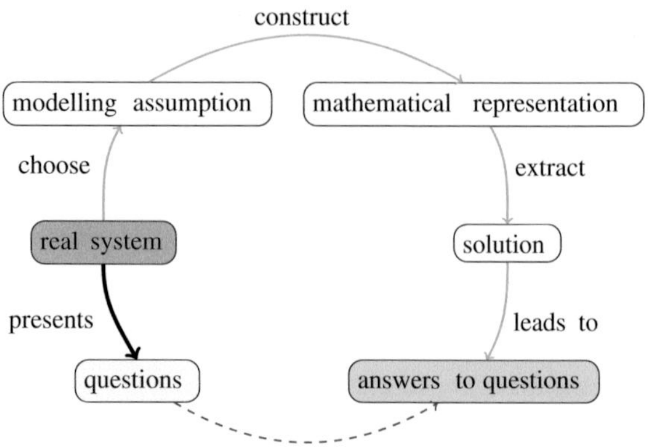

Fig. 1 Constructing mathematical models to answer questions about the world

but rather with things like tennis balls, blocks of concrete, wood studs, steel beams, planets, galaxies, impure water, etc. In the presence of such real systems, we formulate specific questions that determine what aspect of the system is the behaviour of interest. Here are examples of such questions: Would this structure break under a typical load? Would a certain solution containing likely impurities remain stable under a certain increase of temperature? Can the observed trajectory of Uranus be explained by the presence of another heretofore unobserved planet? The task of mathematically modelling real systems is to derive a mathematical representation of the system that will allow us to correctly capture some of its physical properties. From this point of view, a good model does not have to correctly capture all aspects of the system, but only those relevant to the questions that concern us in the first place. Moreover, the question-driven character of applied mathematics makes it clear that representations do not have to be true in order to lead us to correct answers—*selective accuracy* is sufficient.

It is hard to give a completely general account of the way in which mathematical representations are constructed in order to answer our questions about the behaviour of interest, but Fig. 1 perhaps comes close. Starting from a raw, non-mathematical real system, we choose idealized building blocks as our representational medium and attempt to articulate what is mathematically relevant to the problem. The selection of modelling assumptions is a crucial step, which is often plagued with error and uncertainty. It is sometimes possible to say whether modelling assumptions are factual or not, but it is typically hard to directly assess such claims in a comparative way. In other words, it is hard to determine whether one assumption is as far from the truth as another. Making such judgements is even more difficult for sets of modelling assumptions. Moreover, whether this error will invalidate our answers cannot be determined at this point. While we only have particular facts about a system, the evolution of that system is underdetermined. Without an underlying theory providing general kinematic principles, such as a geometrical structure and general conservation laws, no modelling equations could be derived to characterize the behaviour of the system. Similarly, a general theory does not have sufficient specificity to predict anything about the behaviour of systems without being supplemented with specific modelling assumptions [7]. It is this interconnected collection of hypotheses that faces the tribunal of experience.

Indeed, this collection of hypotheses determines dynamical equations characterizing the temporal behaviour of the system (i.e., equations describing the evolution of points or regions in a state space through time). The evolution rule is typically a differential equation (continuous time) or a difference equation (discrete time). Finding the trajectory in the state space prescribed by the rule amounts to solving the system, and it is a very crucial step in extracting the information needed for empirical tests. Without effective solution methods, there is no prediction nor explanation, only speculation. However, this step often involves significant analytical and computational challenges, and we return to this theme in next section. But presuming that information has been accurately extracted, we would then obtain answers to our questions and evaluate the successfulness of our model.

Success as a Balancing Act

The ineliminable need to set aside complexity puts us in a situation in which the sets of modelling assumptions from which we derived model equations are extremely simplified compared to what would faithfully capture real physical systems. Needless to say, when we build a model for a system of real bodies, the inaccuracy and incompleteness of the modelling assumptions could very well lead us to incorrectly answer questions about the behaviour of interest. To establish whether this is the case, a traditional view enjoins the modeller to compare the idealized model to a de-idealized model derived from an accurate and complete set of modelling assumptions. This would allegedly guarantee that the model equations thus derived would correctly answers our questions about the system. Batterman [1] pinned down the idea nicely: "The aim here is to effect a kind of convergence between model and reality. One tries, that is, to arrive at a completely accurate (or 'true') description of the phenomenon of interest. On this view, a model is better the more details of the real phenomenon it is actually able to represent mathematically." However, to experienced applied mathematicians, it is clear that the more details are built into the model, the more mathematically intractable the mathematical equations representing the behaviour of interest are likely to be. That means that, even if we can somehow derive model equations from our accurate and complete set of modelling assumptions, it is likely that we will not be able to use them to make predictions and to obtain answers to our questions concerning the behaviour of interest. That would be a representation that is not manageable, no matter how true, accurate, or complete it is—it would be useless to us.

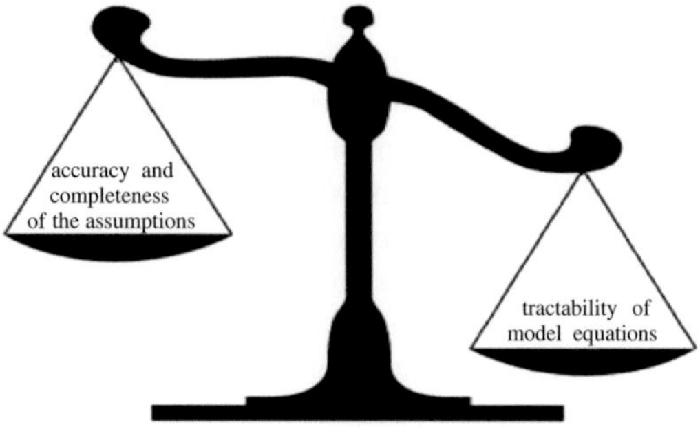

Fig. 2 The fundamental balancing act at the core of applied mathematics

The improvement of the accuracy of the modelling assumptions brings about a decline in the tractability of the model. Thus, counter-balancing the view of the role of applied mathematics as a language for formulating true representations of systems, there is the view that mathematics is "the art of finding problems we can solve," as Hopf said (cited in [2]). Since in applied mathematics there is in addition a question of accuracy, there is always a cost-benefit analysis to perform. The most important contribution of mathematics to modelling is that it provides the tools to do just that. What makes mathematical modelling difficult is that above all we must find a balance between accuracy, completeness, and tractability, as in Fig. 2. Finding this balance with respect to the behaviour of interest is the *true measure of success* in applied mathematics.

But how can we know whether we have reached this balance? How do we distinguish accidental positive results from models that truly capture the essential features of the system? Perhaps this is the mysterious part that will resist our efforts to bring it into the fold of reason.

Rationalizing the Uncanny Accuracy of Mathematics

We have seen that the complexity of the world is such that models will typically not be exact representations of physical systems. Moreover, even simplified models typically have a level of complexity such that extracting information from modelling equations will lead to an additional layer of error. Thus, the key to successful applications of mathematics is to establish that a description of the behaviour of a system is in fact approximately true, *without causing an overflow of information* that would undermine our ability to assess the situation. In this respect, an essential virtue of mathematics is that it can be applied to itself in a way such that *finding whether a representation is close to the truth is easier than finding what the truth is.* I will explain this slogan below based on general insights from perturbation theory and numerical analysis.

The concepts of *sensitivity to* and *robustness under perturbations* play a crucial role in any perspective on error management. There are many rigorously defined concepts in applied mathematics which capture aspects of this very general idea (e.g., Lyapunov exponents, condition numbers, Lipschitz constants, etc.). But for the sake of this essay, an intuitive illustration will suffice. Most of us have at some point had to live in an apartment of questionable quality. Taking a shower in apartments of this ilk is not always without danger. Indeed, a very slight push on the shower knob—technically, we call this a perturbation of the knob's position—might lead to quite dramatic changes in water temperature. In such a case, we can say that the water temperature is very sensitive to perturbations. Robustness under perturbations is just the opposite. If you were so lucky to have air conditioning in this apartment, odds are that however much you cranked the knob, the ambient temperature would not change much. So, the ambient temperature was robust under perturbations. Great accuracy in the shower knob position would be required to correctly predict water temperature, but large errors in the AC unit's knob position could be tolerated in order to accurately predict ambient temperature.

The idealization, simplification, error, and uncertainty contained in models we construct to characterize some behaviour of interest can also be understood as perturbations. Let me first illustrate the point with a modern approach to understanding the impact of computational error as it occurs in computer simulations. Suppose that a dynamical system specified by an ordinary differential equation $x' = f(x)$ and an initial condition $x(0) = x_0$ has been derived as in Fig. 1 to represent a given physical system. Using some computer algorithm, we find a trajectory $\hat{x}(t)$ that will hopefully describe the behaviour of the system accurately. However, there is no a priori guarantee that it will. We need to first analyze the various sources of error. The applied mathematical toolbox offers many ways of talking about error. In what follows I will use the notion of *residual error* as it is easiest to interpret in physical contexts [4]. If we somehow knew the exact solution $x(t)$ to our dynamical system, we would find that $x' - f(x) = 0$ just by re-arranging the terms. However, since the simulated solution $\hat{x}(t)$ contains some degree of computational error, $\hat{x}' - f(\hat{x})$ would not be equal to 0. The quantity Δ given by $\Delta = \hat{x}' - f(\hat{x})$ is, what we call, the residual error. Now, we can reverse-engineer the situation. Instead of saying that $\hat{x}(t)$ is hopefully approximately solving the equation from the dynamical system, we can say that it is an exact solution to the dynamical system $x' = f(x) + \Delta$. With this change of perspective, we can now treat Δ as a perturbation of the original dynamical system. It could be thought of as a breeze, a vibration, a small gravitational effect, or anything relevant. If the magnitude of the computational error, reinterpreted in physical terms, is smaller than the expected modelling error and uncertainty, then the computed solution is deemed *true to our modelling assumptions*.

In fact, for all we know, such a solution could exactly represent the physical system. As mentioned, the representation $x' = f(x)$ is not in general exact due to various sources of modelling and experimental error. However, as the expressive power of mathematics is virtually unlimited, one can presume that some other equation exactly represents the system, say $x' = f(x) + \varepsilon g(x)$, where ε is a small term and $\varepsilon g(x)$ acts as a correcting factor. We could once again verify the residual error of the computed solution $\hat{x}$ mentioned above, but this time with respect to the 'true' equation. Of course, we will not in general have an exact characterization of the correction factor and, as a result, will not exactly know the value of the residual error. However, we can study the amplitude of the residual using *qualitative methods* over various intervals of time, such as the limiting behaviour of the residual error as t goes to infinity. The intervals and parameters of choice for the error analysis will once again be determined by the behaviour of interest. Often, we will find that the residual error is vanishingly small. We then have a precise and effective method to rationalize the fact that even if the construction and manipulation of our successful mathematical models of reality is riddled with uncertainty, measurement error, modelling error, analytical approximations, computational approximations, and other forms of guesses and ignorance, they can be remarkably accurate.

Explaining Miracles Away

The uncanny accuracy of mathematics has been claimed to be miraculous in the sense that it does not seem to admit any rational explanation. To decide whether such a claim can be defended, it is necessary to have a conception of what might be received as a rational explanation. This, in turn, requires a correct understanding of the "logic" of model construction and model assessment. Precisely articulating such metatheoretical concepts is the province of epistemology. As a consequence, a solution to the problem of the applicability of mathematics will be of an epistemological nature, rather than of a metaphysical one.

In addition to the de facto presence of falsehoods, errors (intended and not intended), approximations, and uncertainty (including both known and unknown unknowns) in science, there are other elements that we have not yet mentioned. Indeed, cases of fortunate mistakes, aesthetic preferences, and personal idiosyncrasies of influential figures are also integral parts of real science. However, it does not follow that all those factors play an equally important role in epistemology, as its point is to explain the reliability of scientific knowledge and to delimit its scope. As a result, epistemology does not take the actual thought processes of scientists as its objects, or the actual words used by scientists, or even what scientists take their own activity to be. Rather it envisages a better scenario in which the claims, hypotheses, models, theories, and methods are accounted for not by fortunate mistakes, idiosyncrasies, etc., but by a rationally compelling presentation they ought to have. To use the term introduced by Carnap [3], the object of scientific epistemology is a *rational reconstruction* of science.

The dimension of the rational reconstruction process that generates an object of study suitable for a properly epistemological analysis is often presented as an invective to distinguish the context of discovery from the context of justification. The distinction between the contexts is one between processes of discovery and methods of justifications. The phrase "methods of justification" denotes what satisfactorily establishes knowledge claims, independently of the beliefs of the historical actors.

It is important to emphasize that which methods of justification are rationally admissible is not god-given, as there is room for disagreement. It is nonetheless clear that what is to be included in the context of justification is determined by what methods and tools are considered rational. Different choices might result in different organizations of what belongs to what context. The stakes are clear: if the methods of justification we are willing to admit are too restricted, then some essentially successful scientific practices will appear to be miraculous, without any rational grounds.

Scientists and philosophers alike often depict the scientific method as containing two essential methods of justification. On the one hand, there are the methods of probability theory and statistics which are meant to underly inductive inferences from observed phenomena. On the other hand, logic and axiomatics are meant to capture the deductive structure of scientific theories. Probability and statistics are essentially about making precise judgements about the likelihood of hypotheses. Deductive logic is essentially about truth-preserving inferences (i.e., inferences such that if their premises are true, so will their conclusion). These methods are undoubtedly rationally admissible when properly utilized, but it is essential to emphasize that they do not exhaust the field of rational justifications. It is therefore necessary to revise and supplement our "rational reconstruction toolbox," for otherwise significant parts of applied mathematical sciences would be wrongly considered methodologically unsound.

The successes of applied mathematics crucially depend on the methods of perturbation theory. The type of questions they address are not about probability, likelihood, or truth-preserving inferences. Instead, they concern questions of this type: if causal factors were slightly changed or if parameters were tweaked in various ways, what impact would it have? To see the contrast with deductive logic even more sharply, one could say that the methods of perturbation theory are essentially about determining the circumstances in which arguments with false premises lead to accurate conclusions. Deductive logic cannot address such questions, for even if its inference forms preserve truth, they do not in general preserve approximate truth. Perturbation methods give us the resources we need to learn how to live with falsehood, and this is key to understanding the factors that make so many mathematical models and theories uncannily accurate.

To conclude, persistent failures to unravel the mystery of the applicability can be attributed to an insufficiently rich way of rationally reconstructing scientific and mathematical knowledge. To the extent that the problem of uncanny accuracy is concerned, we need to suitably enrich the catalogue of methods admissible for the rational reconstruction of the concepts of science and mathematics. We should not contemplate elaborate counterfactual constructions about pristine theories that contain no error and uncertainty, but learn how to live with them, and love them, for they are the conditions of possibility of successful science. Then, and only then, the allegedly miraculous character of the applicability of mathematics will be demystified.

References

1. Robert W. Batterman, *Asymptotics and the Role of Minimal Models*, British Journal for the Philosophy of Science **53** (2002), 21–38.
2. Armand Borel, *Mathematics: Art and science*, The Mathematical Intelligencer **5** (1983), no. 4, 9–17.
3. Rudolph Carnap, *The logical structure of the world*, University of California Press, Berkeley, 1928, Translation Rolf A. George, 1967.
4. Robert M. Corless and Nicolas Fillion, *A graduate introduction to numerical methods, from the viewpoint of backward error analysis*, Springer, New York, 2013, 868pp.
5. Paul A. Dirac, The relation between mathematics and physics, The Collected Works of P.A.M. Dirac (1924–1948) (R.H. Dalitz, ed.), Cambridge University Press, 1939, 1995, pp. 122–129.
6. Albert Einstein, *Geometry and experience*, Sidelights on relativity, Dover Publications, 1923.
7. Nicolas Fillion and Robert M. Corless, *On the epistemological analysis of modeling and computational error in the mathematical sciences*, Synthese **191** (2014), 1451–1467.
8. Bertrand Russell, *The art of philosophizing, and other essays*, Philosophical Library Inc, New York, 1968.
9. Clifford Truesdell, *Statistical Mechanics and Continuum Mechanics*, An Idiot's Fugitive Essays on Science, Springer-Verlag, New York, 1980, pp. 72–79.
10. Eugene Wigner, *The Unreasonable Effectiveness of Mathematics in the Natural Sciences*, Communications on Pure and Applied Mathematics **13** (1960), 1–14.

Why Mathematics Works so Well

Noson S. Yanofsky

Abstract A major question in philosophy of science involves the unreasonable effectiveness of mathematics in physics. Why should mathematics, created or discovered, with nothing empirical in mind be so perfectly suited to describe the laws of the physical universe? To answer this we review the well-known fact that the defining properties of the laws of physics are their symmetries. We then show that there are similar symmetries of mathematical facts and that these symmetries are the defining properties of mathematics. By examining the symmetries of physics and mathematics, we show that the effectiveness is actually quite reasonable. In essence, we show that the regularities of physics are a subset of the regularities of mathematics.

Introduction

One of the most interesting problems in philosophy of science and philosophy of mathematics deals with the relationship between the laws of physics and the world of mathematics. Why should mathematics so perfectly describe the workings of the universe? Significant areas of mathematics are formed without anything physical in mind, and yet such mathematics can be used to describe the laws of physics. How are we to understand this?

This mystery is seen most clearly by examining the power of mathematics to determine the existence of physical objects before there is empirical evidence of those objects. One of the more famous examples of the predictive abilities of mathematics is the discovery of Neptune by Urbain Le Verrier simply by making some calculation

Noson S. Yanofsky has a Ph.D. in mathematics (category theory). He is a professor of computer science at Brooklyn College of The City University of New York. In addition to writing research papers he also co-authored "Quantum Computing for Computer Scientists" (Cambridge University Press, 2008). He authored "The Outer Limits of Reason: What Science, Mathematics, and Logic Cannot Tell Us" (MIT Press 2013) which has been received very well both critically and popularly. He lives in Brooklyn with his wife and four children.

N.S. Yanofsky (✉)
Brooklyn College, New York, USA
e-mail: noson@sci.brooklyn.cuny.edu

A. Aguirre et al. (eds.), *Trick or Truth?*, The Frontiers Collection,
DOI 10.1007/978-3-319-27495-9_13

about the abnormalities of the orbit of Uranus. Other examples are P.A.M. Dirac's prediction of the existence of positrons and James Clerk Maxwell's extrapolation that varying electric or magnetic fields should generate propagating waves.

Even more amazing is that there existed entire established fields of mathematics long before physicists realized that they were useful for understanding various aspects of the physical universe. The conic sections studied by Apollonius in ancient Greece were used by Johannes Kepler in the beginning of the seventeenth century to understand the orbits of the planets. Complex numbers were invented several centuries before physicists started using them to describe quantum mechanics. Non-Euclidian geometry was developed decades before it was used in an essential way for general relativity. (Details of these and other remarkable mathematical discoveries can be found in [5].)

Why should mathematics be so good at describing the world? Of all thoughts, ideas, or ways of expressing things, why should mathematics work so well? What about other modes of thought? Why does poetry fail to describe the exact movements of the celestial bodies? Why can't music capture the full complexity of the periodic table? Why is meditation not very helpful in predicting the outcomes of experiments in quantum mechanics?

The problem of why mathematics works so well was famously addressed by Nobel prize winning physicist Eugene Wigner in a paper titled "The unreasonable effectiveness of mathematics in the natural sciences" [4]. Wigner did not arrive at any definitive answers to the questions. He wrote that "the enormous usefulness of mathematics in the natural sciences is something bordering on the mysterious and … there is no rational explanation for it."

Albert Einstein described the mystery succinctly:

> How can it be that mathematics, being after all a product of human thought which is independent of experience, is so admirably appropriate to the objects of reality? Is human reason, then, without experience, merely by taking thought, able to fathom the properties of real things? [1]

To be clear, the problem really arises when one considers both physics and mathematics to each be perfectly formed, objective and independent of human observers. With such a conception, one can ask why these two independent disciplines harmonize so well. Why can an independently discovered law of physics perfectly be described by (already discovered) mathematics?

Many researchers have pondered this mystery and offered various solutions to the problem. Theologians solve the mystery by positing a Being who perfectly set up the laws of the universe and used the language of mathematics to describe these laws. However the existence of such a Being adds to the mystery of the universe. Platonists (and their cousins Realists) believe that there exists some realm of "perfect Forms" which contains all mathematical objects, structures and truths. In addition, this "Platonic attic" also contains all the physical laws. The problem with Platonism is that, in order to explain the relationship between our mathematical world and the physical universe, it invokes yet another Platonic world. Now one must explain the relationship between all three of these worlds. Other questions also arise: are imperfect mathematical theorems also a perfect Form? Do outdated laws of physics also reside in Plato's attic? Who set up this world of perfect Forms?

The reason most cited (see [2]) to answer the unreasonable effectiveness is that we learn mathematics by examining the physical universe. We understand some of the properties of addition and multiplication by counting stones and sheep. We learned geometry by looking at physical shapes. From this point of view, it is not a surprise that physics follows mathematics since the mathematics that we know was formed by scrutinizing the physical world. The main problem with this solution is that mathematics is very useful in areas distant from human perception. Why is the hidden world of subatomic particles so perfectly described by a mathematics learned from observing stones and sheep? Why is special relativity, which deals with objects that move near the speed of light, described by a mathematics that was learned by watching objects that move at normal speeds?

While these and other purported solutions to the unreasonable effectiveness of mathematics have some merit, the mystery still remains. A deficiency in most of the discussions is that it is assumed that everyone already knows exactly what mathematics is. The true nature of mathematics is usually taken as so obvious that it is not even discussed. In our discussion we take aim to rectify this by giving an exact definition of mathematics. In two recent papers [6, 7], Mark Zelcer and I have formulated a novel view of the nature of mathematics. We show that just as symmetry plays an important defining role in physics, so too, symmetry plays an important defining role in mathematics. This view of mathematics supplies an original solution to the unreasonable effectiveness problem.

This essay is organized as follows. First we recall how symmetry plays a major role in modern physics. We then go on to discuss the importance of symmetry in mathematics. The next part brings these two sections together to explain why physical laws are naturally expressed in a mathematical language. We close the essay with some thoughts on some deeper questions.

What is Physics?

Before we can contemplate the reason why mathematics describes physical laws so well, we have to be exact as to the definition of a physical law. To say that a physical law describes some physical phenomenon is a bit naïve. It is much more. As a first attempt we might say that each law describes *many* different physical phenomena. For example, the law of gravity describes what happens when I drop my spoon; it describes what happened when I dropped my spoon yesterday; and it describes what happens if I drop my spoon on the planet Saturn next month. A law of physics describes a whole bundle of many different phenomena. However this definition is not enough. A single physical phenomenon can be perceived in many different ways. One can perceive a phenomenon while remaining stationary at the same time as another perceives the same phenomenon while moving (in a uniform constant frame, or an accelerating frame). Physics states that no matter how a single phenomenon is perceived, it should be described by a single physical law. We conclude that a physical law describes a whole bundle of *perceived* physical phenomena. For example, the law of gravity describes my observation of a spoon falling in a speeding car while I am in the car; a stationary friend's observation of the spoon falling in the speeding car; an observation by someone standing on his head near a black hole of the falling spoon in the speeding car on Saturn, etc. Every physical law describes a large bundle of perceived physical phenomena. That single law should be able to describe all the diverse perceived physical phenomena.

The question then arises as how to classify all perceived physical phenomena into different laws or which perceived physical phenomena should be bound up together as one. When do two perceived physical phenomena really represent the same physical law? Physicists use the notion of symmetry for this. Colloquially the word "symmetry" is used to describe physical objects. We say a room has symmetry if the left side of the room is the same as the right side of the room. In other words, if we swap the furnishings from one side to the other, the room would look the same. Scientists have extended this definition of symmetry to describe physical laws. A physical law is symmetric with respect to a type of translation if the law still describes the translated phenomenon. For example, physical laws are symmetric with respect to location. This means that if an experiment is done in Pisa or Princeton, the results of the experiment are the same. The phenomenon occurring in Pisa is bundled up with the phenomenon occurring in Princeton. Physical laws are also symmetric with respect to time, i.e., performing the same experiment today or tomorrow should give us the same result. In terms of bundles, that means that all experiments performed at any time are bundled up to be in the same class of perceived physical phenomena. Another obvious symmetry is orientation. If you change the orientation of an experiment, the results of the experiment remain the same. We use both the languages of bundles of perceived physical phenomena and of symmetries of physical laws.

There are many other types of symmetries that physical laws have to obey. Galilean relativity demands that the laws of motion remain unchanged if a phenomenon is observed while stationary or moving at a uniform, constant velocity. Special relativity states that the laws of motion must remain the same even if the observers are moving close to the speed of light. General relativity states that the laws are invariant even if the observer is moving in an accelerating frame.

Physicists have generalized the notion of symmetry in many different ways: gauge transformations, local symmetries, global symmetries, continuous symmetries, discrete symmetries, etc. Victor Stenger [3] unites the many different types of symmetries under what he calls *point of view invariance*. That is, all the laws of physics must remain the same regardless of how they are viewed. He demonstrates how much—but not all—of modern physics can be recast as laws that satisfy point of view invariance. This means that different perceived physical phenomena are bundled together if they are related to the same physical phenomenon but are perceived from different points of view.

The real importance of symmetry came when Einstein formulated the laws of special relativity. Prior to him, one first found a law of nature and then found its symmetries. In contrast, Einstein used the symmetries to discover the laws. In order to find the laws of special relativity, he posited that the laws must be the same for a stationary observer and an observer moving close to speed of light. Given this presupposition, he went on to formulate the equations that describe special relativity. This was revolutionary. Einstein had realized that symmetries are the defining characteristics of laws of physics. Before Einstein physicists would say

A physical law satisfies symmetries.

After Einstein, they say

Whatever satisfies symmetries is a physical law.

To elaborate, Einstein's revolution was the realization that a physicists selects from the many different perceived physical phenomena those that have symmetry. Those perceived physical phenomena that exhibit symmetries are bundled together and called a law of physics.

In 1918, Emmy Noether showed that symmetry is even more central to physics. She proved a celebrated theorem that connected symmetry to conservation laws that permeate physics. The theorem states that for every symmetry of a certain type there exists a conservation law and vice versa. For example, the fact that the laws of physics are invariant with respect to translations in space corresponds to conservation of linear momentum. Time invariance corresponds to conservation of energy. Orientation invariance corresponds to conservation of angular momentum. Equipped with the understanding given by Einstein and Noether of the centrality of symmetry, physicists have been searching for novel and different types of symmetries in order to find new laws of physics.

With this broad definition of physical law, it is not hard to see why the laws have a feeling of being objective, timeless and independent of human observations. Since the laws are applied in every place, at every time, and from every perspective, they have a feeling of being "out there." However, we can look at it another way. Rather than saying that we are looking at many different instances of an external physical law, we may say that we humans select those perceived physical phenomena that have some type of regularity and bundle them together to form a single physical law. We act like a sieve that picks and chooses from all the physical phenomena that we perceive. We bundle together what is the same physical law, and we ignore the rest. We cannot eliminate the human element in understanding the laws of nature.

From the bundle point of view, one can understand unification of physical laws. Newton's insight was that the same law of physics that describes celestial mechanics also describes terrestrial mechanics. In other words, the phenomenon of the moon being pulled towards the earth can be bundled together with the phenomenon of an apple being tugged towards the earth. James Clerk Maxwell showed that the laws of electricity and magnetism can be bundled into one law we now call electromagnetism. Steven Weinberg, Abdus Salam and Sheldon Lee Glashow showed that the seemingly separate bundles of phenomena that correspond to the electromagnetic force can be bundled up with the weak force to form the single electroweak force.

Before we proceed, we must mention a symmetry that is so obvious it has not been articulated. A law of physics must satisfy *symmetry of applicability*. That is, if a law works for a particular physical object of a certain type then it will work for any other physical object of the same type. For example, if a law is true for one positively charged subatomic particle moving close to the speed of light, then it will work for another positively charged subatomic particle that is also moving close to the speed of light. In contrast, that law might not work for a macroscopic object moving at a slow speed. The perceived physical phenomena for all these objects will be bundled together as one law. Symmetry of applicability will be of fundamental importance when we discuss the relationship of physics to mathematics.

What is Mathematics?

Let us spend a few minutes considering the real essence of mathematics. We illustrate with several examples.

Long ago a farmer realized that if you take nine apples and combine them with four apples, there will be thirteen apples. Not long after that it was noticed that if nine oranges are combined with four oranges, there will be thirteen oranges. That is, if you swap every apple for an orange, the amount of fruit remains the same. At some point an early mathematician looked at many instances of this and bundled them together to summarize with the mathematical expression $9 + 4 = 13$. This pithy little statement encapsulates all the instances of this type of combination. The expression will be true for any whole discrete object that can be exchanged for apples.

Topologists discuss the *Jorden Curve Theorem.* This says that any closed (starts and finishes at the same point) simple (non-self-intersecting) curve on a plane splits the plane into an "interior" region and an "exterior" region. No matter how large or complicated the curve is, the theorem says that there will always be two separate regions. This works for every closed simple curve. If we change one curve for another curve, we will get another two different regions.

On a more sophisticated level, a major theorem in algebraic geometry is *Hilbert's Nullstellensatz* which is essential for understanding the relationship between ideals in polynomial rings and algebraic sets. For every ideal J in a polynomial ring there is a related algebraic set, $V(J)$, and for every algebraic set S there is an ideal I(S). The relationship between these two operations is given as follows: for every ideal J we have $I(V(J)) = \sqrt{J}$ where $\sqrt{J}$ is the radical of the ideal. If we change one ideal for another ideal, we get a different algebraic set. If we swap one algebraic set for another, we get a different ideal. Essentially, the bundling together of *all* the different instances of this law is Hilbert's theorem.

One of the basic ideas in algebraic topology is the *Hurewicz homomorphism*. For any topological space X and positive integer k there exists a group homomorphism from the kth homotopy group to the kth homology group $h_* : \pi_k(X) \to H_k(X)$. This homomorphism has special properties depending on the space X and the integer k. If the space X is swapped for space Y and k is exchanged for k' there will be another group homomorphism $\pi_{k'}(Y) \to H_{k'}(Y)$. Once again, no single instance of the statement has any mathematical content. Rather, it is the realization that *all* the instances of the statement can be bundled together that makes this mathematics.

In these examples, we focused on changing the semantics of the mathematical statements. We exchanged oranges for apples. We switched one closed simple curve for another closed simple curve. We swapped one ideal for another. We replaced one topological space for a different one. Our main point is that when you make the appropriate changes, the mathematical facts remain true. We claim that this ability to alter the semantics of a mathematical statement is the defining property of mathematics. That is, a statement is mathematics if we can swap what it refers to and it remains true.

Associated with every mathematical statement is a class of entities called the *domain of discourse* for that statement. The statement is stating something about all the elements of this domain of discourse. When a mathematician says "For any integer n...," "Take a Hausdorff space...," or "Let C be a cocommutative coassociative coalgebra with an involution...," she is setting up the domain of discourse for that statement. If the statement is true for some element in the domain of discourse, it is true for any other. Notice that the associated domain of discourse for a single statement can consist of many types of entities.

With the concept of domains of discourse in mind one can see why the uses of variables are central to mathematical practice. Variables are placeholders in mathematical expressions that tell how to transform referents in statements. A variable indicates the type of object that is being dealt with and the way to change its value within the statement. In Hilbert's Nullstellensatz, the variable J stands for any ideal. In the Hurewicz theorem example, X was a topological space and k was a natural number. We can change X for any topological space and k for any number.

This swapping of one element in the domain of discourse for another can be seen as a type of symmetry. If you swap one referent for another referent within the domain of discourse, the fact will remain true. We call this *symmetry of semantics*. Mathematics is invariant with respect to symmetry of semantics. We are claiming that this symmetry is as fundamental to mathematics as symmetry is to physics. In the same way that physicists formulate laws, mathematicians formulate mathematical statements by determining which bundle of ideas satisfies symmetry of semantics. Mathematicians would say that

Mathematical statements satisfy symmetry of semantics.

But they should now say

Any statement that satisfies symmetry of semantics is mathematics.

Logicians will find symmetry of semantics to be a familiar notion. They call a logical statement *valid* if it is true for every element of the domain of discourse. Here we are saying that a statement satisfies symmetry of semantics if we can exchange any element of its domain of discourse with any other element of its domain of discourse. The novelty we are expressing here is that validity can be seen as a form of symmetry and that this symmetry is the defining feature of mathematics.

One might object to this definition of mathematics as being too broad. They would say that what we are defining is any general statement. The argument would be that a statement that satisfies symmetry of semantics is not just a mathematical statement but is a general statement. There is a two-prong defense to this criticism. First, modern mathematics is very broad. Mathematics is not just about numbers and quantities. Looking at modern mathematics one finds shapes, propositions, sets, categories, microstates, macrostates, qualities, etc. In order to deal with all these objects, our definition of mathematics needs to be broad. A second defense is that there are many general statements that do not satisfy symmetry of semantics. "It is cold in New York during the month of January," "Flowers are red and green," and "Senators are honest people" are general statements but do not satisfy symmetry of semantics and hence are not mathematical. As long as there are any counterexamples to such statements within their implied domain of discourse, they may be general, but they are not mathematical. The fact that symmetry of semantics does not permit any counterexamples within the domain of discourse implies a certain precision of thought and language which people associate with mathematics.

Mathematical statements also have other types of symmetries that they satisfy. A simple example is *symmetry of syntax*. This says that a mathematical object can be described (syntax) in many different ways. For example we can write 6 as 2×3, or $2 + 2 + 2$ or $54 \div 9$. Similarly we can talk about a "non-self-intersecting continuous loop," "a simple closed curve," or "a Jordan curve" and mean the same thing. Our point is that the results of the mathematics will be the same regardless of which syntax is used. In practice mathematicians tend to use the simplest syntax possible, like "6" instead of "$5 + 2 - 1$".

Other symmetries that mathematical statements possess are so obvious and taken for granted that even mentioning them seems strange. For example, mathematical truths are invariant with respect to time and space: if they are true now then they will also be true tomorrow and they were true before human beings even existed. If they are true in Albany they are true on Alpha Centauri. It is similarly immaterial if the mathematical truth is asserted by Mother Teresa or by Oswald Teichmüller. No one cares where, when, or in what language a theorem is stated. It is even irrelevant if the mathematical statement was stated at all.

With mathematics satisfying all of these different types of symmetries, it is easy to see why mathematics—like physics—also has the feel of being objective, timeless, and independent of human observers. Since the facts of mathematics apply to many different objects, are discovered by many different individuals working independently, and in many different times and places, one can start believing that math is somehow "out there." But, we need not make that leap. Symmetry of semantics is at the core of how we determine mathematical truths. Human beings function like sieves that pick and choose from among thoughts and ideas. We bundle the thoughts that are related by symmetry of semantics and declare such statements to be mathematics. We do not say that there exists some perfect mathematical truths and we humans find many different instances of that truth. Rather, we say that there are many different instances of a mathematical fact and we humans bundle them together to form a clear mathematical statement.

From this symmetry formulation of mathematics, it is easy to see what the concept of unification in mathematics is all about. When you have two distinct fields of mathematics and they are shown to be intimately related in a way that results of one field can be used to get results of another field, you have a type of unification. An example of such unification is monstrous moonshine. This subject describes the shocking connection between the monster group and modular functions. Another example is the Langlands program which connects Galois groups in algebraic number theory to automorphic forms and representation theory of algebraic groups. The Erlangen program can also be seen as a way of unifying different types of geometries using group theory. In all these examples of unifications, there is a function (or an isomorphism) from the domain of discourse of one field to the domain of discourse of another field. The symmetries of one field (the true mathematical statements) will than map to symmetries of the other field. Category theory is an entire branch of mathematics that was created to describe such unifications. The founders of category theory invented a language that was based on algebraic topology, which is a branch of mathematics that unifies algebra and topology. Category theory is now used in many areas to show that seemingly different parts of mathematics (and physics and theoretical computer science) are closely related.

See [6, 7] for much more discussion about the nature of mathematics and how other issues in the philosophy of mathematics are dealt with from this perspective.

Why Mathematics Works Well at Describing Physics?

Armed with this understanding of the nature of physics and mathematics we can tackle the question of why mathematics works so well at expressing physical laws. Let us look at three physical laws.

Our first example is gravity. A description of a single instance of a perceived physical phenomenon of gravity might look like this: "On the second floor of 5775 Main Street in Brooklyn, New York at 9:17:54 PM, I saw an 8.46 ounce spoon fall and hit the floor 1.38 s later." While it might be totally accurate, it is not very useful and it is not the description of all the instances of the law of gravity. As we explained, a law consists of all the perceived instances of that law. The only way to capture all of the bundled perceptions of physical phenomena of a particular law is to write it in mathematical language which has all its instances bundled with it. Only Newton's formula $F = G\frac{m_1 m_2}{d^2}$ can capture the entire bundle of perceived physical phenomena of gravity. By substituting the mass of one body into m_1 and the mass of the other body into m_2 and the distance into d, we are describing an instance of gravity.

In a similar vein, to find the extremum of an action one needs to use the Euler–Lagrange equations: $\frac{\partial L}{\partial q} = \frac{d}{dt}\frac{\partial L}{\partial \dot{q}}$. The symmetries of an action and its local maximums or minimums can be expressed with these equations that are defined with the symmetry of semantics. Of course, this can also be expressed by a formula that uses other variables or other symbols. By symmetry of syntax, we can even write the formula in Esperanto. It is irrelevant how and in what language the mathematics is expressed. As long as it is mathematics because only mathematics can capture the action principle in all its instances.

The only way to truly encapsulate the relationship between pressure, volume, number of moles and temperature of an ideal gas is the ideal gas law: $PV = nRT$. This small mathematical statement has all the instances of that law built into it.

In the three examples given, all the perceived instances of the physical laws can only be expressed with a mathematical formula. The perceived physical phenomena that we are trying to express are all implied by all the instances which are inherent in the mathematical statement. In terms of symmetry, we are saying that the physical symmetry of applicability is a special type of the mathematical symmetry of semantics. In detail, for any physical law, symmetry of applicability states that the law can deal with swapping any appropriate object for any other appropriate object. If there is a mathematical statement that can describe this physical law, then we can substitute different values for the different objects that one is applying. In terms of bundles, we are saying that every bundle of perceived physical phenomena is a sub-bundle of instances of the mathematical law that describes it.

The point we are making is that mathematics works so well at describing laws of physics because they were both formed in the same way. The laws of physics are not living in some Platonic attic nor are the central ideas of mathematics. Both the physicist and the mathematician chose their statements to be applicable in many different contexts. We bundle perceived physical phenomena in the same way we bundle instances of mathematical truth. It is not a mystery that the abstract laws of physics are stated in the abstract language of mathematics. Rather the regularities of phenomena and thoughts are seen and chosen by human beings in the same way. The fact that some of the mathematics could have been formulated long before the law of physics is discovered is not so strange since they were formed with the same notion of symmetry.

We have not completely solved the mystery concerning the unreasonable effectiveness of mathematics. There are still deep questions lurking here. For one, we can ask why humans even have physics or mathematics. Why do we notice regularities or symmetries of physical phenomena or of thought? The answer is that part of being alive is being somewhat homeostatic, that is, living creatures must act to preserve themselves. The more they comprehend of their environment, the better off they will be. Inanimate objects like sticks and stones do not react to their environment. In contrast, plants turn towards the sun and their roots search for water. As living creatures become more sophisticated they notice more about their environment. Human beings notice many regularities of the world around them. While chimpanzees do not seem to understand abstract algebra and clever dolphins do not write textbooks in quantum field theory, humans do have the ability to grasp the regularities of their perceived physical phenomena and the thoughts that go through their head. We call the regularities of our thoughts "mathematics" and some of these regularities manifest themselves as regularities of perceived physical phenomena, which we call "physics."

From this perspective, we can understand the usual answer to Wigner's unreasonable effectiveness. Many people say that the reason why mathematics is so effective is that we learn mathematics from the physical world. While that is not totally true, we can say that many times the symmetries or regularities of the physical world have taught us some symmetries and regularities of the mathematical world.[1]

On the deepest level one can ask why there are any regularities at all in perceived physical phenomena? Why should it be that an experiment performed in Poughkeepsie will yield the same results as if it was performed in Piscataway? Why should balls roll down ramps at the same speed even if they are released at different times? Why should chemical reactions be the same when they are perceived by different people and in different ways? To answer such questions, we appeal to the anthropic principle. This is a type of reasoning that formulates answers from the very fact that we exist. If the universe did not have some regularities, no life would be possible. Life uses the fact that the physical universe contains some repeating patterns. Since there is life in the universe, there has to be certain regularities in the laws of physics. If the universe was totally random or like a psychedelic vision, no life—in particular no intelligent human life—would survive. Anthropic reasoning does not eliminate the question. It tells us that the fact that we are here means that the universe must be a certain way, and had the universe been another way, we could not exist to ask the question. But the anthropic principle does not tell us *why* the universe is that way. We are still left with those deep—and, as yet, unanswerable—questions of "Why is the universe here?" "Why is there something rather than nothing?" and "What's going on here?"

The previous paragraph is a partial answer to why there are any regularities in perceived physical phenomena. Now let us ask why there are any regularities of thought called mathematics. First notice that the physical universe would go on perfectly well without any human being understanding the mathematics behind the laws of physics. It is not hard to think of some crazed politician starting global thermonuclear war and ending all life on Earth. After such a catastrophic event, the laws of nature will go on their merry way without the slightest pause or hiccup. So the existence of physics does not depend on the existence of mathematics. We are left with the question of why there exists any mathematics at all.

Perhaps we can borrow a trick that physicists use to explain why the universe is fine-tuned for intelligent life to help us understand why there is any regularity of thought called mathematics. Some physicists looking at the physical world say that the universe is, in fact, *not* fine-tuned for intelligent life. Everywhere they look in the universe they do not see intelligent life. There are billions of billons of stars each with many planets which seem to be devoid of any life. The Fermi paradox asks "Where is everyone?" If the universe is so fine-tuned for life, why have extraterrestrials neither been seen nor stopped by for tea. One—of many—answers to the Fermi paradox is

---

[1] Sometimes what we learn from the physical world might be an error. The world around us looks continuous. Space looks continuous and time feels continuous. In order to talk about space, time, and space-time, we use the continuous real numbers. However modern physicists tell us that space, time and space-time are not continuous. There is a minimum distance called Plank length and a minimum duration called Plank time. In a sense, the real numbers were invented by mistake. Perhaps if we were much smaller and were better aware of the discreteness of time and space we would have never come to invent the real numbers. One can only speculate on such matters.

that ET is not there. Of the billions and billions of planets, none (or very few) other than ours has developed intelligent life forms. Even in our planet there are large parts that are underwater, too hot, too cold, or too dry to support intelligent life forms. So rather than saying that the universe is fine-tuned for intelligent life, these physicists assert that the vast majority of our universe is not fine-tuned for intelligent life. They say that a miniscule percentage of the universe is fine-tuned for intelligent life. This takes away some of the mystery of the so called fine-tuned universe. We can perhaps say the same thing about mathematics. We asked why there is any regularity of thought which we call mathematics. We can answer that the vast majority of thought does not, in fact, show any regularity. Most thought has counterexamples and does not have the precision that is necessary to be called mathematics. Only a miniscule percentage of thought has any regularity. We humans just pick out minute parts of thought that does have regularity and call that mathematics. This takes away some of the mystery of the existence of mathematics.

While we have not eliminated all the mysteries, we have shown that any existing structure in our perceived physical universe is naturally expressed in the language of mathematics.

References

1. Einstein, Albert. "Geometry and experience." http://www.relativitycalculator.com/pdfs/einstein_geometry_and_experience_1921.pdf, 1921.
2. Ronald E. Mickens, editor, *Mathematics and Science*, World Scientific, Singapore, 1990.
3. Victor Stenger. *The Comprehensible Cosmos: Where Do the Laws of Physics Come From?* Prometheus Books, Amherst, NY, 2006.
4. Eugene P. Wigner. "The unreasonable effectiveness of mathematics in the natural sciences." Communications in Pure and Applied Mathematics, 13(1), February 1960.
5. Noson S. Yanofsky. *The Outer Limits of Reason: What Science, Mathematics, and Logic Cannot Tell Us*, MIT Press, 2013.
6. Noson S. Yanofsky and Mark Zelcer, "Mathematics via Symmetry". http://arxiv.org/pdf/1306.4235v1.pdf, 2013.
7. Noson S. Yanofsky and Mark Zelcer, "The Role of Symmetry in Mathematics." http://arxiv.org/pdf/1502.07803v1.pdf, 2015.

Genesis of a Pythagorean Universe

Alexey Burov and Lev Burov

Abstract The full-blown multiverse hypothesis, chaosogenesis, is refuted on the grounds of simplicity, the large scale and high precision of the already discovered laws of nature. A selection principle is required not only to explain the possibility of life and consciousness, but also theoretizability of our universe. The anthropic principle provides the former, but not the latter. As chaosogenesis is shown to be the only thinkable scientific answer to the question of why the laws of nature are the way they are, its refutation means that this question cannot be answered scientifically.

Introduction

The task of science, as it is generally assumed, is to find the laws of nature allowing both to explain the diversity of observations as well as to predict new ones. Science seeks to discover the logic that is hidden beneath phenomena and which determines their flow and qualities. The understanding of truth as uncovering of hidden essence, as dis-covery, is embedded in the Greek word α λ ή θ ε ι α (truth), consisting of negation (α-) and λ ή θ η, which means a veil or concealment. Pythagoras taught that this essence is the harmony of hidden unity which can be expressed in the language of numbers. When Galileo stated that nature is a book written in the language of mathematics, he was expressing this ancient Pythagorean credo. The same can be said about Dirac, whose fundamental belief was that "the laws of nature should be expressed in beautiful equations", and about Einstein who believed that the strongest and noblest motive for the scientific search is a deep conviction of the rationality of the universe, saturated with the cosmic religious feeling.

When theories that exhaust phenomena are formulated and logically unified into a single theory of everything, the task of fundamental science is finished. Whatever this theory of everything may be, other theories in physics will be its consequences as

A. Burov (✉)
FNAL, Batavia, IL, USA
e-mail: burov@fnal.gov

L. Burov
Scientific Humanities LLC, San Francisco, CA, USA

A. Aguirre et al. (eds.), *Trick or Truth?*, The Frontiers Collection,
DOI 10.1007/978-3-319-27495-9_14

limit cases or asymptotes. Although humanity does not now and may possibly never have such a theory in its fullness, many of its limit cases are known to us as concrete theories, such as classical and quantum mechanics, general theory of relativity, the standard model, and others.

The laws of nature are discovered as composite and specific mathematical structures. As these structures are revealed, we unavoidably come to a certain question regarding the structures themselves. First of all, why does any law expressed by one or another mathematical formula structure our world at all? While it is thinkable for a universe to be structured by any logically consistent system, out of this infinite set of structures only one determines our universe. Why this structure and not another? Why are the laws simple enough to be discovered? Why are they mathematically beautiful? Who or what singled them out and on what ground?

In this way the laws of nature become a problem, though not in the usual scientific context of searching them out, but as something that requires its own explanation. The illusory nature of an explanation that does not go beyond natural laws was pointed out by Ludwig Wittgenstein [1]:

> The whole modern conception of the world is founded on the illusion that the so-called laws of nature are the explanations of natural phenomena. Thus people today stop at the laws of nature, treating them as something inviolable, just as God and Fate were treated in past ages. And in fact both are right and both wrong: though the view of the ancients is clearer in so far as they have a clear and acknowledged terminus, while the modern system tries to make it look as if everything were explained.

Here Wittgenstein criticizes a silent acceptance of a composite and special mathematical structure as the ultimate explanation of the world. Such explanation barred from further questioning and not subject to reasonable ground of its own existence is an affirmation of unreasonableness of this ground.

So what is this 'unreasonableness' that stops reason in this questioning? Declaring the unreasonable and meaningless, which can not even be questioned, as a foundation of reason is nothing but an assertion of the primacy of the absurd. We define the absurd as a derogation of reason, a denial of its independent value, which should not be confused with related, but different entities: foolishness, meaninglessness, impossibility, the fantastic, humorous or accidental. Foolishness is but a weakness, a lack of ability to reason. To point out foolishness, to sneer at it, only emphasizes the significance of thinking. A magic carpet is fantastical and perpetual motion impossible, but notions about them in no way debase the significance of reason; if anything, they can stir minds to new, daring pursuits. In play, theater, and literature, making fun of common sense and challenging it constitutes an absurdist joke, and, more generally, an absurdist genre. But this challenge to reason is limited by play and the space of imagination. There is no meaning in chance, but it is thinkable; often randomness can be accounted for probabilistically, and in some situations its presence can be diminished through thoughtful consideration. Chance itself is unreasonable, but it doesn't touch upon values and doesn't put reason under an existential threat. That which does deliver such an attack, is the absurd.

In the context of the limits of reason, the antithesis of absurd is mystery. Mystery is reason's creative source, literally its alma mater that is replete with meaning. While both are presented as limitations, where the absurd is a dead end, mystery is infinity. Acceptance of the power of the absurd is a fundamental denigration of reason; yet the awareness of mystery ennobles and inspires rational thought.

Paul Davis, in considering the denial of reasonableness of the question about the source of the laws of nature, characterizes it specifically as the assertion of fundamental absurdity [2]:

> One can ask: Why that unified theory rather than some other?... Why a unified theory that permits sentient beings who can observe the moon? One answer you may be given is that there is no reason: the unified theory must simply be treated as "the right one," and its consistency with the existence of a moon, or of living observers, is dismissed as an inconsequential fluke. If that is so, then the unified theory—the very basis for all physical reality—itself exists for no reason at all. Anything that exists reasonlessly is by definition absurd. So we are asked to accept that the mighty edifice of scientific rationality—indeed, the very mathematical order of the universe—is ultimately rooted in absurdity!

Such superstition destroys the meaning of fundamental science by undermining the importance of reason, subjected by this superstition to the absurd.

So then, what could be the answers concerning the source of the laws of nature? Is there any way of choosing or rejecting one or another? That is the topic of discussion in the present article.

The Fine Tuning Question

"There is now broad agreement among physicists and cosmologists", writes Paul Davis [3], "that the universe is in several respects 'fine-tuned' for life". Similarly, Stephen Hawking has noted:

> The laws of science, as we know them at present, contain many fundamental numbers, like the size of the electric charge of the electron [*fine structure constant*] and the ratio of the masses of the proton and the electron. ... The remarkable fact is that the values of these numbers seem to have been very finely adjusted to make possible the development of life [4].

Another crucial point is articulated by Alexei Tsvelik [5]:

> [since] the number of existing life-imposing conditions by far exceeds the number of constants, their fulfillment could not be achieved by fine tuning of these constants and required also the right choice of the fundamental principles of physical laws.

The premise of the fine-tuned universe revived the old metaphysical problem of the source of order in the world as the problem of fine-tuning: who or what tuned the universe so fine? A pure scientific approach required finding an objective answer: not "somebody" but "something" as the cause of tuning.

Order from Chaos

It is thought that this "something" could be any combination of laws of nature provided by one or another general theory and random factors; or, using the terms of Platonic philosophy, any combination of forms and chaos. However, as it was noted by Wittgenstein, any theory used in that respect itself requires to be explained. John A. Wheeler expressed the same as a question: why is this very theory structuring everything existent? Why doesn't some other theory instead? In other words, the use of any theory for this does not solve the problem of fine-tuning, but moves it to a higher level. The only way to solve this problem totally in the framework of science is to show a possibility of appearance of being from nothing, or *chaosogenesis*, the appearance of order from chaos. Indeed, theories, being specific formal structures, are limited and composite entities, and thus lead to the question "why this theory and not other?". Chaos per se is limitless and structureless, a totality intrinsically undivided into "this" and "that", whose various manifestations differ from each other due only to the variety of doors that one or another theory opens for chaos to enter. Historically, the idea of chaosogenesis is very old, having been traced down to Hesiod and pre-Socratics, and it had been opposed by the Pythagoreans and Platonics. For instance, Plotinus wrote: "Any attempt to derive order, reason, or the directing soul from the unordered motion of atoms or elements is absurd and impossible." Jaspers [6] Not all contemporary cosmologists share Plotinus' views on the chaosogenesis, so the idea is frequently pronounced.

Max Tegmark has formulated the "Ultimate ensemble theory of everything", whose main motivation is clearly expressed [7]:

> If the TOE [*theory of everything*] exists and is one day discovered, then an embarrassing question remains, as emphasized by John Archibald Wheeler: Why these particular equations, not others? Could there really be a fundamental, unexplained ontological asymmetry built into the very heart of reality, splitting mathematical structures into two classes, those with and without physical existence? After all, a mathematical structure is not "created" and doesn't exist "somewhere". It just exists. As a way out of this philosophical conundrum, I have suggested that complete mathematical democracy holds: that mathematical existence and physical existence are equivalent, so that all mathematical structures have the same ontological status.

Thus for Tegmark, the terminus ultimately explaining everything existing is the totality of all mathematical forms, the Platonic world. To "just exist", the mathematical structure has to be self-consistent, logically acceptable, but what he doesn't mention is the unity of these forms. This unity must not only somehow bind every one of them together but it has to guarantee their self-consistency. The forms though are mental entities. They are not thinkable without a mind which contains them as truly self-consistent. Thus, we have to conclude that this unity, the terminus of Tegmark's questioning, is a mind, even if it is not mentioned at all. What makes this mind special and distinctive from its various Platonic versions is its total indifference to the forms it contains. That is what Tegmark calls "the mathematical democracy."

It has to be stressed, that purely by itself, without any forms involved, chaos cannot produce anything, and Tegmark's model is not an exception to this rule:

it assumes that all possible worlds are based on mathematical structures, such as groups, algebras, fields, sets of equations, and other formal systems. It also assumes that there is a way for these structures to show themselves as phenomena, and to be observed both as mathematical and physical objects. Chaos comes in this picture as the randomness of a universe we happen to be born in, with the only limitation that the laws of this universe are compatible with life and consciousness. What makes Tegmark's model very special is its minimal involvement of a priori concretization or selection principles, which is why we are equating this model of "mathematical democracy" with chaosogenesis.

A possibility for the structure of the fundamental laws of nature to be random to some unclear degree and beyond that to be non-randomly selected by some unpronounced entity was expressed by several leading scientists, e.g. by Andrei Linde (see a citation below) and Steven Weinberg [8]:

> ...we have to keep in mind the possibility that what we now call the laws of nature and the constants of nature are accidental features of the big bang in which we happen to find ourselves, though constrained (as is the distance of the Earth from the Sun) by the requirement that they have to be in a range that allows the appearance of beings that can ask why they are what they are.

The Darwinian theory of evolution is widely believed to explain the birth of order from chaos. To follow its line of thought, our universe is considered a member of a huge or infinite ensemble of universes, one generated by the other, with daughter universes mostly inheriting the logical structure of the mother ones, adding some mutations on top [9, 10]. After the heredity and variation of the multiplying logical structures are settled, the third Darwinian principle, selection, can be introduced as well. This role is played by the so called weak anthropic principle, or WAP [11], pointing out that only those universes can be observed where observers can appear, which selects a narrow class of fine-tuned universes as it is noted in Weinberg's quotation above. Thus, though our universe is thought of in this Darwinian approach as a random representative of Tegmark's totality of forms, its fine tuning apparently receives a scientific explanation as a result of a Darwinian chaosogenesis. Although in the infinite megaverse only a tiny portion of universes is fine-tuned for life and consciousness, the probability for any observer to see the universe as fine-tuned is one hundred percent.

An important role of WAP as the only alternative to theistic explanations of the fine tuning was stressed by Weinberg [12]:

> In me, this apparent fine-tuning arouses wonder. The only explanation for it, other than a theological explanation, is in terms of a multiverse—I mean a universe consisting of many parts, each with different laws of nature and different values for its constants, like the 'cosmological constant' which governs cosmic expansion. If there is a multiverse consisting of many universes, most of them hostile to life but a few favorable to it, then it's not surprising that we find ourselves in one where conditions are in the fortunate range.

Nothing seemingly contradicts the assumption that our universe is a random representative of WAP-selected subset of Tegmark's multiverse, but is that really so?

Does the universe indeed have no clear signature excluding any possibility of it having been randomly selected from this totality of all possible mathematical structures? Is the concept of chaosogenesis irrefutable by any thinkable observation, i.e. is it not a scientific hypothesis? Apparently, it is considered as irrefutable by some leading experts. For instance, Brian Greene clearly states just that [13]:

> I draw the line at ideas that have no possibility of being confronted meaningfully by experiment or observation, not because of human frailty or technological hurdles, but because of the proposals' inherent nature. Of the multiverses we've considered, only the full-blown version of the Ultimate Multiverse falls into this netherland. If absolutely every possible universe is included, then no matter what we measure or observe, the Ultimate Multiverse [*i.e. Tegmark's one*] will nod and embrace our result.

Contrary to B. Greene, we are showing below that Tegmark's hypothesis runs counter to certain observations, so it fails, and fails as a scientific theory.

Weak Anthropic Principle

On the question of possibility of the long evolution from the Big Bang to thought the WAP answers thus: in those worlds, where this path hasn't been traversed, there is no one to ask. But then in our universe this path has been traversed, so we ask one more question: why isn't the path thrown into nowhere right now? Why does this world not only exist, but continues to exist, and the prediction of its continued existence comes true over and over, while the prediction of the end of the world turns out to be false again and again? What keeps this complex world with its life and thought in being?

The prediction of an immediate end of the world is completely unavoidable in the framework of the WAP and full-blown multiverse. Maintaining whatever special features is a special requirement, demanded of the universe. Special demands can be fulfilled, if appropriately grounded. If there is no ground, then there is no sense in expecting of keeping the requirements. The WAP explains why life and thought became possible. But out of the truism, which it uses to explain, no logical consequence follows that further on the required conditions will remain satisfied.

The reader might ask, if it already turned out this way with our universe, that up till now it maintained life, does it mean that it has some kind of a foundation of the laws that it happened to have, which keep it in this status of continuation of life. What's wrong with this explanation of the renewed anthropic continuation?

Let's consider this explanation more closely. It supposes an existence of some laws, giving structure to the universe, its evolution in time. The laws themselves at the same time must be atemporal: otherwise whatever segregation of them from the temporal world would be meaningless; the regulators would be no different than regulated. But even beside that they are atemporal, the laws are mental entities: to see them and to think them is one and the same. Postulating laws as objective mental entities implies Mind as a sphere of their being. It takes nothing but a Mathematical Mind to differentiate a law good for the universe from one which is not—meaninglessness, absurdity or a self-contradictory system. Because the Mind not only discerns the laws from non-laws but it manifests them as structure-forming elements of the material universes, It is also a Maker. In this way, the very assumption of some non-contradictory laws leads to the conclusion about the transcendental Creator as the Mathematical Mind and Maker, even strictly within the framework of Tegmark's full-blown multiverse.

Let's assume that Tegmark's multiverse is just that ocean, a random drop of which is our universe; a chance limited by the WAP. Does this assumption mean that the conditions for life, satisfied for billions of years will continue to be satisfied in the next second? Does the belonging to the full-blown multiverse, strengthened by billions of years of good behavior serve as the ground to conclude that in the next second this behavior will continue to be good? There is no such ground here. If a mathematical function of a general form, a random representative of all possible functions, has been at zero so far, then we can only conclude that this specific quality will not continue to be maintained even in the very near future.

Tegmark's multiverse, determined by all non-contradictory sets of formulas, essentially is no different from a multiverse limited by nothing, that is pure chaos. Whatever the behavior of the universe up till now, there will always be an infinite number of laws corresponding to this behavior, and the chance to select out of them the set of laws that guarantees good behavior even for the next second equals to zero. There is an infinite number of laws of explosive action in Tegmark's multiverse, sleeping up to a certain moment and waking arbitrarily soon. To exclude the awakening of every one of this infinity of infinitely complex laws in the next second would equate to postulating an ungrounded specificity of our universe within the multiverse.

And so, just the conformity of the universe to smooth laws is not enough to conclude its good behavior in the nearest future. The unavoidable conclusion on this basis is the immediate end of the world. In order to avoid this conclusion, it is necessary to rule out laws of explosive action from the initial multiverse, because the truism of WAP does not exclude them. Essentially, induction logic is ungrounded for Tegmark's multiverse, and this point was noted by Leslie and Kuhn [14]:

> What is meant, though, by "all mathematical structures"? Leibnitz wrote in his Discourse on Metaphysics that no matter how you scatter dots on paper there will be some mathematical formula, perhaps tremendously long and complicated, that generates a line passing through every one of them... Absolutely any universe, no matter how disorderly, might therefore count as having "a mathematical structure"... Well, what if the dots already scattered looked "very orderly" through lying in a straight line so that a simple formula fitted them? Countless far more complicated formulas would fit those dots as well... So if he lives in a multiverse containing absolutely all mathematical structures, shouldn't Tegmark expect that the line of his future would wriggle wildly? That he would almost surely become a pile of dust or a goldfish or a cupcake or... or... or... or...?

What is left then of the original motivation of "not needing this hypothesis," of the Creator, the motivation responsible for the WAP? The above analysis of implications demonstrates complete failure of that plan. According to these conclusions, which we come to unavoidably, the Ultimate Mind is necessary not only as a Mathematical one, guaranteeing non-contradiction of laws, but also as an Architectural one, limiting participation of laws of explosive action. It will be shown further that the laws of our universe point to yet one more substantial selection.

A Cosmic Observer

"Observers" in WAP are not normally specified; it is not taken into account what it is namely they do observe. We suppose that to be qualified as observers they at least have to be conscious, as it is also reasonable to assume that conscious creatures observe their immediate space of life support and have access to at least empirical knowledge about it. However, this sort of knowledge has nothing to do with theoretical knowledge of the big cosmos; the first by no means entails the second. Let us fix this point of an important distinction, a distinction between those *simple, minimal, empirical observers* and *cosmic observers*, who are discovering theories of big cosmos, seeing their universe both at extremely large and extremely small scales, far exceeding the scale of immediate life support. To become cosmic observers, minimal ones must live in a very specific world among the populated worlds. Specifically, their universe has to be theoretically comprehensible on a big cosmic scale; their world has to be *theoretizable*, so to say. In other words, the possibility for observers to be not just simple but cosmic requires their universe to have a very special logical structure: it has to be described by elegant laws, covering many orders of magnitude of their parameters. Contemporary humanity is indeed a cosmic observer. For today, our scale of scientific cognition is described by an enormous dimensionless parameter $\sim 10^{45}$; that big is the ratio of the sizes of largest object of physics, the universe, $\sim 10^{26}$m, to the smallest ones, the top quark and the Higgs boson, corresponding to $\sim 10^{-19}$m.

The Condition of Elegance

This condition of *theoretizability* apparently is extraneous to the selective anthropic principle, that is, theoretizability seems unnecessary for the universes to be populated by conscious creatures or to be observed. In fact, the latter condition is essentially local; it requires something like a life-friendly planet inside any universe. The former condition, though, is global; it requires the laws of nature to be elegant on a big cosmic scale, a scale by far exceeding that of the life on the planet. Generally, local conditions do not entail global consequences, and since theoretizability is a specific functional requirement detached from WAP selection, we have to conclude that it is highly unlikely for an observed universe to be theoretizable. That is how R. Collins puts it [15]:

> …even if we assume that embodied observers could only arise in a region with simple laws, the problem still remains: for every simple law of some form X, there is an infinite number of complex laws that have form X when approximated to the conditions obtaining in small spatiotemporal region around X but are extremely complex everywhere else in space and time.

Since we know, after Isaac Newton, that our universe is theoretizable, chaosogenesis theory is apparently refuted. However, this refutation, being qualitative only, leaves a possibility to object. Its core statement, that theoretizability is a specific requirement *detached* from the anthropic condition for universes "to be observed" can be questioned. How can we be sure that theoretizability is logically independent from WAP? It would not be independent, if WAP did not allow for our theoretizable laws of nature any visible modifications even at extremes of very large and very small scales, modifications that might exclude the appearance of conscious beings for one or another reason. The very concept of a fine-tuned universe is suggesting to us that sort of an idea concerning the fundamental constants, and so, we may ask: what if the same is true concerning the very structure of the laws of nature? Although it would be hard to believe that moderate modification of, say, General Relativity at the distances exceeding the solar system, can dramatically reduce the possibility of consciousness on our planet, we should consider a chance that it cannot be excluded. This very argument for the strong relation between the weak anthropic principle and theoretizability was recently suggested by Linde [16]:

> ... the inflationary multiverse consists of myriads of 'universes' with all possible laws of physics and mathematics operating in each of them. We can only live in those universes where the laws of physics allow our existence, which requires making reliable predictions.

The same idea was expressed in the latest book of Tegmark [17], with reference to E. Wigner:

> An anthropic-selection effect may be at work as well: as pointed out by Wigner himself, the existence of observers able to spot regularities in the world around them probably requires symmetries, so given that we're observers, we should expect to find ourselves in a highly symmetric mathematical structure. For example, imagine trying to make sense of a world where experiments were never repeatable because their outcome depended on exactly where and when you performed them. If dropped rocks sometimes fell down, sometimes fell up and sometimes fell sideways, and everything else around us similarly behaved in a seemingly random way, without any discernible patterns or regularities, then there might have been no point in evolving a brain.

Let's accept this arguable hypothesis, and suppose that somehow WAP does not allow significant deviations of the laws of nature locally compatible with conscious beings from the globally theoretizable form. Then, the question is: which deviations from the existing laws are allowed by the anthropic principle? If the world is generated by chaos, all imaginable additional terms to the life-selected ones are coming into play; the amplitude or width of the resulted deviation is limited by the anthropic principle, but functional behavior of the deviation is arbitrary. We have some estimations about the allowed deviation in the context of fine-tuning as relative variations of the fundamental constants compatible with WAP, and the most stringent of them are at the order of 10^{-3}, i.e. 0.1 % [11]. Since we are considering here the problem of functional accuracy, the enormously stringent requirements on some constants, like the initial conditions at the big bang [18], do not reduce the amplitude of these functional variations. Thus, working on the Linde argument, we may roughly estimate the sensitivity of the anthropic selection to the relative functional variations of the fundamental laws to not be finer than 0.1 % or so. If the laws of nature were generated by a random choice from Tegmark's multiverse, they would be expressed by irregular functions possibly following elegant ones within a relative width of ~0.1 % or more. In this respect, it does not matter whether chaos reveals itself through arbitrary functions or arbitrary mathematical structures; with Tegmark's "mathematical democracy" functional representatives of the two families are indistinguishable and are dominated by extremely complicated, practically irregular functions. The elegant formulas might be approximations to the real irregular fundamental laws with that WAP-determined accuracy, but not better.

Moreover, measurements of the fundamental constants in this world would be reproducible only at the anthropic level, not better. If physicists of that hypothetical world tried making measurements of their fundamental constants at the better accuracy, they would realize that none of the measurements are reproducible at that level; they would all contain space-time noise with a relative amplitude of 0.001, driven by infinitely complicated terms of the true laws of nature. So, physics in that Tegmarkian universe would be stopped at the anthropic accuracy level simply because, with the probability of 100 %, no reproducible measurement would be possible there with accuracy better than that.

We know though, that the real accuracy of our fundamental theories is not only better than anthropic, but many orders of magnitude better; they are absolutely precise on that scale. Indeed, the General Relativity test with a double neutron star PSR 1913+16 showed an unprecedented agreement between theory and observation at the level of 10^{-14}. Another impressive demonstration of extremely high precision relates to the Quantum Electrodynamics: the theoretically predicted value of an electron's magnetic moment is confirmed by measurements with the accuracy $\sim 10^{-11}$; see e.g. Ref. [18]. Thus, many experiments which proved high precision of our elegant laws of nature, orders of magnitude better than the anthropic width, show that the laws of nature cannot be selected from all mathematical structures by anthropic requirements only; their simplicity cannot be a consequence of their anthropic character.

This consideration shows, by the way, that cosmological chaosogenesis is a scientific hypothesis since it is falsified by observations. Note that the idea of the multiverse was at least partly motivated by the wish to find a pure scientific explanation to the fact of the fine-tuned universe: if our universe is the only one, its fine-tuning does not suggest any other reasonable explanation but an act of purposeful creation. For a single universe, its fine-tuning is too stringent for a purely scientific explanation, but the idea of multiverse chaosogenesis, suggested as an attempt to explain fine tuning within bounds of science, is refuted by the opposite reason: the estimated anthropic limitations on fine-tuning aren't anywhere fine enough to explain the experimental confirmations of the extreme precision of the elegant forms as fundamental laws.

A Pythagorean Universe

After having announced the "complete mathematical democracy" at the beginning of his article, later on Tegmark notices that "our physical laws appear relatively simple". At this point, to be consistent with reality, he gives up the proclaimed "mathematical democracy" in favor of an aristocracy of simple mathematical forms. After such an overturn, his multiverse now has almost nothing to do with chaos; instead, it is generated by some source of elegant mathematical forms. As a result, "the embarrassing question" about the source of this ontological inequality of the mathematical forms remains as it was. This contradiction of Tegmark's "democratic" intention with his "aristocratic" practice was noted by Alex Vilenkin [19]:

> Tegmark's proposal, however, faces a formidable problem. The number of mathematical structures increases with increasing complexity, suggesting that "typical" structures should be horrendously large and cumbersome. This seems to be in conflict with the simplicity and beauty of the theories describing our world.

Since the laws of our universe are not picked randomly, they can only be purposefully chosen. Our universe is special not only because it is populated by living and conscious beings but also because it is theoretizable by means of elegant mathematical forms, both rather simple in presentation and extremely rich in consequences. To allow life and consciousness, the mathematical structure of laws has to be complex enough so as to be able to generate rich families of material structures. From the other side, the laws have to be simple enough to be discoverable by the appearing conscious beings. To satisfy both conditions, the laws must be just right. The laws of nature are fine-tuned not only with respect to the anthropic principle but to be discoverable as well. Multiple aspects of this double fine tuning are discussed in Ref. [20] and references therein. In other words, the Universe is fine-tuned with respect to what can be called as the *Cosmic Anthropic Principle*: its laws are purposefully chosen for the universe to be *cosmically observed.* It could be even that our laws belong to the simplest of possible sets that permit our sort of life. Would it be possible to take any part away from our existing theories without compromising forms of life as we know them? Such a special universe deserves a proper term, and we do not see a better choice than to call it *Cosmos or* to qualify it as *Pythagorean*, in honor of the first prophet of theoretical cognition, who coined such important words as *cosmos* (order), *philosophy* (love of wisdom), and *theory* (contemplation).

Since chaosogenesis, being limited only by the anthropic principle, is the only option for a completely scientific solution of the problem of cosmogenesis, its refutation entails that the problem of cosmogenesis cannot be solved within the framework of science. Any scientific approach to that would require a specific set of axioms, consistent not only with the anthropic principle but with elegant mathematical forms truly underlying our world; however, we've shown that the question about embedment of one instead of another specific set as a logical structure of the universe cannot be scientifically answered.

Starting with Pythagoras, it was a matter of faith for sparse groups of few people and lonely individuals that "fundamental laws of nature are described by beautiful equations." Theoretical science was conceived and nurtured by this very faith with its "cosmic religious feeling", which inspired scientific cognition for twenty-five centuries. Without any exaggeration, all great theories, from those of Copernicus, Kepler and Newton to those of Einstein and Dirac happened as guesses on the grounds of some fundamentally simple ideas like symmetry, conservation, or equivalence. Likewise, Wigner saw "the appropriateness of the language of mathematics for the formulation of the laws of physics" as a miracle and "a wonderful gift which we neither understand nor deserve" [21]. His maxim "we should be grateful for it" can only have meaning if a mind to be grateful to is implied.

The noted forty-five orders of magnitude of scientific cognition, with more than ten digits of precision reached in some experimental verifications, allow us to conclude about a scientific confirmation of what was considered a matter of faith for two and a half millennia: now it is a matter of fact that the universe is indeed Pythagorean. In other words, the existence of the Platonic world of elegant mathematical forms structuring the physical world is scientifically confirmed, and the accuracy of this confirmation is many orders of magnitude better than that of any specific statement of physics.

After two and a half millennia since its birth, fundamental science reached a grade of maturity allowing for a dual confirmation of its faith: the Pythagorean faith is confirmed as prophecy coming true and as a good tree that brings forth good fruit.

Three Worlds, Two Totalities

Pythagorean forms of the discovered laws of nature tell us that the ultimate goal of fundamental physics, the theory of everything, either contains a significant Pythagorean core, or, what is more reasonable to assume, is totally Pythagorean. This Pythagorean core has to be powerful enough to generate a sufficiently rich set of Pythagorean laws, as we observe, but whatever this theory of everything is, it cannot be the ultimate answer to the question about the order of being, because this form is special due to there being other forms, and so, like any other, it does not constitute a totality. For laws of nature, there are only two thinkable explanatory principles, opposites of each other, which are totalities: chaos and mind as such.

Because the logical structure of our universe can not be explained by chaos, and because it can not explain itself, we are left with only one possible explanation remaining, that it was conceived and realized by a mind. A. Vilenkin prefers to formulate this apparently inevitable conclusion about the cosmic Mind as a question [19]:

> … the laws should be "there" even prior to the universe itself. Does this mean that the laws are not mere descriptions of reality and can have an independent existence of their own? In the absence of space, time, and matter, what tablets could they be written upon? The laws are expressed in the form of mathematical equations. If the medium of mathematics is the mind, does this mean that mind should predate the universe?

To be a complete terminus of questioning, a creative mind has to be mind per se, or the Absolute Mind. Otherwise, questions about origin and possibility of its

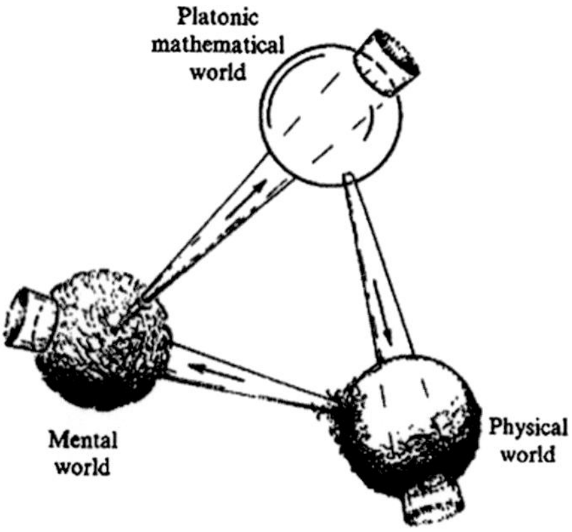

Fig. 1 Roger Penrose, "Three Worlds, Three Mysteries" [18]

mindness would require new answers. Unlike chaos, Absolute Mind as terminus leaves room for mystery; the creativity of the human mind does as well. Where there is mystery, questioning is inexhaustible, and the feeling of mystery instills a deep value in the pursuit of knowledge. Contrary to this, the postulation of chaosogenesis, by rejecting the primacy of mind, is incompatible with mystery, and so with the value of fundamental cognition. Thus, the problem of cosmogenesis leads to a dual mystery, one aspect of which is the Absolute Mind as the source of the laws of nature, while the other aspect lies in a mind capable of discovering them. From this point of view, Tegmark's multiverse obtains a new meaning; it is a space for the search for interesting worlds to be created, with laws open to discovery.

It seems important to mention here that chaos, refuted as a possible source of the laws of nature, can and does participate in the physical world as indeterminism, by means of uncertainty left by the quantum laws of nature.

The very idea of observation, being so far associated with material objects only, is enriched by an even more fundamental meaning of the Platonic observation, i.e. observation of elements of the Platonic world structuring the material world. Cosmic observation is possible only due to a combined vision of both worlds. Roger Penrose suggested the idea and the image of "Three Worlds, Three Mysteries" (Fig. 1) [18]. The three worlds, Physical, Platonic, and Mental, differ time-wise. The Platonic world does not have any age at all; it is atemporal. The Physical world is temporal, and its age, counted from the border of all observations, the big bang, is calculated at 13.798 ± 0.037 billion years (note the precision!). The age of humanity as cosmic observers is extremely short on that scale. Yet although the history of many scientific discoveries is known minutely, and although we cannot observe anything closer than our thoughts, the genesis of the cosmic observer remains no less mysterious to us than the genesis of Physical and Platonic worlds.

Wonder of Pythagorean harmony of the fundamental laws of nature and continuing demonstrations of human ability to discover them so remotely from our own natural scale leads now more than ever before to deep questions about the three mysteries, whose entanglement and coherence are revealing the underlying Unity, the ultimate transcendental Source of everything existent, including ourselves, the growing cosmic observers.

The authors are thankful to Alexei Tsvelik and Mikhail Arkadev for stimulating discussions.

References

1. L. Wittgenstein, "Tractatus Logico-Philosophicus" (1921).
2. P. Davies, "The Goldilocks Enigma", Houghton Mifflin Harcourt. Kindle Edition (2006).
3. P. Davies, "How bio-friendly is the universe". Int. J. Astrobiol 2, p. 115 (2003).
4. S. Hawking, "A Brief History of Time", Bantam Books, p. 125 (1988).
5. A. Tsvelik, "Life in the Impossible World", Ivan Limbakh publ., St.-Petersburg, 2012 (in Russian).
6. K. Jaspers, "The Great Philosophers", A Harvest Book, Vol. 2 (1966).
7. M. Tegmark, "The Mathematical Universe", Foundations of Physics 38 (2), p. 101 (2007).
8. S. Weinberg, "Lake Views: This World and the Universe" (Kindle Locations 337–338). Harvard University Press. Kindle Edition (2011).
9. L. Smolin, "The Life of the Cosmos" (1997).
10. A. Linde, "The Self-Reproducing Inflationary Universe", Sci. Am., Vol. 271, No. 5, pp 48–55, (1994)
11. J.D. Barrow, F.J. Tipler "The Anthropic Cosmological Principle." (1988).
12. J. Holt, "Why Does the World Exist?: An Existential Detective Story" (pp. 156–157). Liveright. Kindle Edition.
13. B. Greene, "The Hidden Reality: Parallel Universes and the Deep Laws of the Cosmos" (Kindle Locations 5943–5947). Knopf Doubleday Publishing Group. Kindle Edition (2011).
14. J. Leslie & R.L. Kuhn, ed. "The Mystery of Existence. Why Is There Anything At All?" (2013).
15. R. Collins, "God And the Laws of Nature", Philo, Vol. 12, No 2, p. 18 (2010)
16. A. Linde, " Why Is Our World Comprehensible?", in "This Explains Everything", Ed. J. Brockman, 2012.
17. M. Tegmark, "Our Mathematical Universe: My Quest for the Ultimate Nature of Reality", Knopf Doubleday Publishing Group. Kindle Edition.
18. R. Penrose, "The Road to Reality", First Vintage Books Edition, 2007.
19. A. Vilenkin, "Many Worlds in One: The Search for Other Universes" (Kindle Locations 3188–3191). Farrar, Straus and Giroux. Kindle Edition (2006).
20. G. Gonzalez and J. W. Richards, "The Privileged Planet", Regnery Publishing, Inc (2004).
21. E. Wigner, "The Unreasonable Effectiveness of Mathematics in the Natural Sciences", Communications on Pure and Applied Mathematics, vol. 13, issue 1, pp. 1–14 (1960).

Beyond Math

Sophía Magnúsdóttir

Abstract In this essay I reflect on the use and usefulness of mathematics from the perspective of a pragmatic physicist. I first classify the different ways we presently think of the relation between our observations and mathematics. Then I explain how we can do physics without using math—that we are in fact already doing it. In the end the pragmatic reader will know why math is reasonably effective, why we are all models, and how to go beyond math.

The Pragmatic Physicist

Once upon a time "universe" meant all there is. But now that physicists have several variants of multiverses the universe isn't what it used to be, and unfortunately no two people agree exactly what the multiverse is either. Latin lovers all over the world are grinding their teeth but it is rather pointless to insist using a word according to its etymology if this just results in communication failure. I will therefore refer to the all-there-is as "All."

For the purpose of this essay I take on the perspective of a strictly pragmatic physicist. The pragmatic physicist—first name Pragmatic, last name Physicist—wants to describe observations and only bothers to think if thinking seems useful for this description.

Pragmatic Physicist doesn't have reason to believe that humans are unique and their observations are special. If he or she[1] refers to observations in general, he thus means all observations that could be done by anybody or anything anywhere at any time. We denote these observations with capital O. If he refers only to observations made by humans so far, we denote these as a small o.

[1] In the hope that the female reader will forgive me, I henceforth use the male article.

S. Magnúsdóttir (✉)
Gothenburg, Sweden
e-mail: sophia.magnusdottir@gmail.com

A. Aguirre et al. (eds.), *Trick or Truth?*, The Frontiers Collection,
DOI 10.1007/978-3-319-27495-9_15

Pragmatic Physicist is interested in the All only to the extent that it concerns either an observation or a tool to describe an observation. He does not know what it might possibly mean for something to "exist" or to "be real" if it can't be observed and not be used for anything, but then he doesn't care enough to deny its existence either. Even leaving aside superstition and religion, the belief that human consciousness has a non-physical component is widely spread. Pragmatic Physicist rolls his eyes upon such romantic interpretations of synapses and neurons, but since it doesn't matter to him, he just ignores the question whether the All is more than that what can be observed and what is being used to describe the observations.

This means we will not discuss here the question what is real and what this might possibly mean, because the pragmatic physicist doesn't care. We will only discuss observations and their description. You are free to believe there is another part of the All and call the pragmatic physicist a soulless moron.

The pragmatic physicist is not a follower of the shut-up-and-calculate doctrine. He knows that usefulness depends on both context and time, and he thinks it is shortsighted to just dismiss philosophy. Like the CEO who has an eye on customers too young to have purchasing power, Pragmatic Physicist has an eye on philosophers' discourse so he is ready if they become useful. If you push him, he admits he can't name any philosopher who makes sense to him, but he also knows that he knows nothing and vaguely recalls some philosopher figured that out a long time ago.

Pragmatic Physicist's view on science is shaped by what he learned as student. The purpose of scientific inquiry in his opinion is to develop models that explain observations. Once a model is found it can often be used to manipulate nature to better suit human needs. This is only possible if the same model is suitable for similar but different subsystems of the observable part of the All. With 'subsystem' Pragmatist loosely means anything that is not the whole (of all observations). These subsystems often fulfill additional properties because otherwise it becomes very difficult to find models, but for our purposes we will not have to specify these properties.

Science works by identifying a model that can be mapped to subsystems of what Lee Smolin likes to call the "Real World Out There" [1] and what Pragmatist refers to as the totality of all possible observations, O. He doesn't really know how to explain operationally what an observation is, but he recognizes one if he sees it. If you insist, he would say that an observation has taken place if one subsystem has obtained information about another subsystem.

In modern physics, models are mathematical. The difference between pure mathematics and physics (and some other parts of the natural sciences) is that a physical theory does not consist solely of mathematics, it also must contain a prescription to identify the mathematical structure with observation. Pragmatist was taught that science is all about finding regularities. These regularities are what the mathematical models capture, and their understanding and subsequent application is what makes science so useful.

To be useful, a theory must do more than reproduce observations on one subsystem—this could be done already by recording the observation. To be useful, a theory must provide an advantage over just waiting and seeing what happens.

It must either make a prediction (about future observations, not necessarily about future events), or succeed in describing many observations by the same explanation.

Pragmatist is aware that he does not use the word "theory" to mean what the US American popular science media tries to convince its readership it means, namely that scientists who say "theory" mean a model that is mapped to observation which has already been found to be correct to high accuracy. The reason he isn't using the word this way is that as a matter of fact it's not how physicists use it, though string theorists certainly wish it was. He is too pragmatic to worry that some Creationist will read an FQXi essay, so Pragmatist just asks you to understand he uses the word "theory" to mean a model including a map to observations, regardless of whether this theory has already been shown to be useful.

The Mathematical Universe

Pragmatic Physicist has read Max Tegmark's paper "The Mathematical Universe" [2]. Tegmark argues that the All is identical to mathematics. Pragmatic finds that mildly interesting but useless and forgets about Tegmark until he comes across Sabine Hossenfelder's blog. Hossenfelder explains she dislikes Tegmark's Mathematical Universe because it unimaginatively assumes that mathematics is the best way to describe observation, rather than just the best way humans have found so far [3].

Now Pragmatist worries. What if Hossenfelder is right and there are other, possibly better, ways to describe observations than using math? What if some of the observations we presently have cannot be described by math at all? What could possibly be better than math? He cannot imagine. Neither, for that matter, can Hossenfelder as she admits in her writing.

On his birthday Pragmatic logs in for his annual facebook visit to upthumb the birthday greetings. He doesn't really know what facebook is good for, but he quite likes the video of the baby armadillo and some hours later he finds himself reading Mark Twain's hilarious essay about "The Awful German Language" [4]. Twain writes "When a German gets his hands on an adjective, he declines it, and keeps on declining it until the common sense is all declined out of it. It is as bad as Latin."

In a flash of insight it occurs to Pragmatic that he just learned something about a language he doesn't speak by reading about it in a different language. Indeed, he now realizes that one could write essays about mathematics using plain old English! While one will never learn German without using German vocabulary, and never learn mathematics without using equations, one can still learn something about one language using a different language.

And so Pragmatic Physicist takes his mathematical toolbox and sets out to apply it to the question whether all observations can be described by math.

A Mathological Classification

Pragmatic physicist reasons: Either Tegmark is right and observations can be mathematics rather than just being described by mathematics, or Tegmark is wrong and this isn't so. In the first case, either all observations are math, or not all. In the latter case, either all observations can be described by math or not. We denote with M the entirety of all mathematics[2] and with O all observations. We know that there exist observations that can be described by math, so if observations are not themselves math there exists some map, T, that maps the models, at least some of which are math, to observations. This leads to the following four cases:

1. Observations can be math and all observations are math $\{O\} \cap \{M\} \neq \{\emptyset\} \wedge \{O\} \cap \{M\} = \{O\}$
2. Observations can be math and not all of them are math $\{O\} \cap \{M\} \neq \{\emptyset\} \wedge \{O\} \cap \{M\} \neq \{O\}$
3. Observations are described by math but are not math and all observations can be described by math $\{O\} \cap \{M\} = \{\emptyset\} \wedge \{T(O)\} \cup \{M\} \neq \emptyset \wedge \{T(O)\} \cap \{M\} = \{T(O)\}$
4. Observations are described by math but are not math and not all observations can be described by math $\{O\} \cap \{M\} = \emptyset \wedge \{T(O)\} \cup \{M\} \neq \{\emptyset\} \wedge \{T(O)\} \cap \{M\} \neq \{T(O)\}$

These four cases can be illustrated in Venn diagrams (Fig. 1). For clarity we have further separately depicted the cases in which subsets are identical to the mother set. We have also in the diagrams added {o}, the observations that humans already have so far, which is a subset of {O}, and their images {T(O)} and {T(o)}. The diagrams show the situation in which some of our observations cannot be described by math, or are not math respectively. The arrows indicate how theories in physics presently operate.

In more detail:

1a. All observations are math, and there is no mathematics that cannot be observed. This is Tegmark's Mathematical Universe whose philosophy we will call Tegmarxism. Science in this case is the task of finding similar mathematical structures and maps between them.

1b. All observations are math but only some of math appears as observation. This sounds much like Garrett Lisi's assertion that yes, the universe is mathematics, but only the prettiest mathematics. We will call this the Mauritian Variant of Tegmarxism.

2a. Observations can be math but some of our observations are deprived of the mathematical ideal form. We will call this the Platonic Street View because it is what the person on the street would think if you told them that math is real.

2b. All math can be an observation, but not all observations are mathematical. This is the extended version of Tegmarxism that Hossenfelder argued for. We will refer to this as Post-Tegmarxism.

---

[2] We will turn to the question whether M contains itself later.

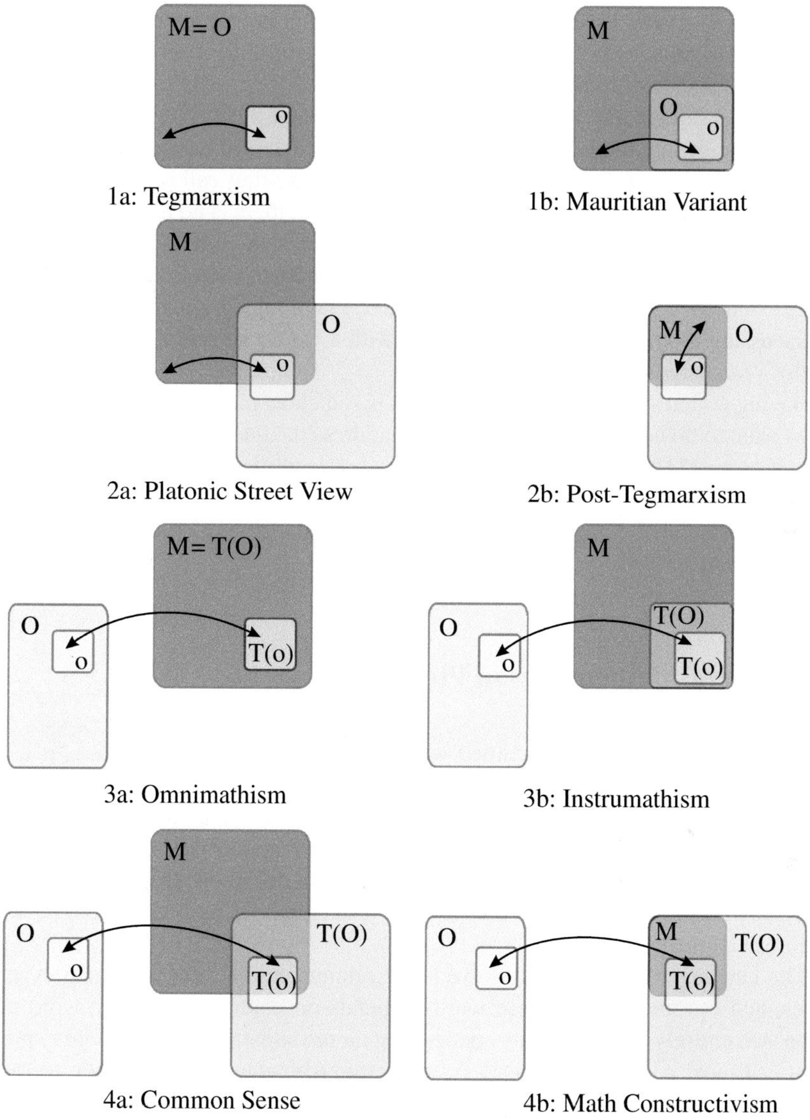

Fig. 1 Classification of relations between mathematics, M, and observations, O. Which one is your philosophy?

3a. All observations can be described by math and all math is used for some observation. In this case Pragmatic Physicist need not worry that his math will ever run out of explanatory power or that some awesome math book might never be put to work. We will call this Omnimathism.

3b. All observations can be described by math, but not all math is necessary. In this case mathematics is the instrument scientists can use to describe all of nature. We will call this Instrumathism.

4a. Observations are described by math but are not math. Not all observations can be described by math and not all math is useful to describe observations. This is what the layperson typically thinks of mathematics, we thus call it Common Sense.

4b. Not all observations are described by math, but there is no math that does not describe some observation. This is a quite untypical view. It is most similar to the philosophy of those arguing that mathematics is a human invention and only exists to the extent that we use it. As explained earlier, Pragmatic Physicist it not interested in discussing the meaning of existence but with a hat tip he refers to this option as Math Constructivism.

Having gotten so far, Pragmatic Physicist is somehow pleased but now he doesn't know what to do next. How can he possibly address the question which one of these cases is correct? How could he find out? And what do we do if mathematics cannot be used to describe all our observations? Searching for inspiration, he turns to Eugene Wigner's influential essay on the "Unreasonable Efficiency of Mathematics in the Natural Sciences" [5].

The Role of Mathematics in Physics

Pragmatic Physicist is not concerned with the question what mathematics "is". He doesn't even know what that question means. He is only concerned with the question what mathematics is good for. Mathematics is often described as a language or a tool, which is also how Wigner refers to the role of mathematics in physics. The difference to other languages and tools—the relevant difference that makes math so dramatically useful—is that it is entirely self-referential. Or, as Tegmark put it, "free of human baggage."

The languages that humans evolved for communication refer to human experiences, and the meaning of these words depends on context. The same word can mean two entirely different things depending on the situation or even facial expression, and most people consciously or unconsciously adapt their language to local customs. Human language is efficient to connect with others around us, not to be precise and reproducible. If you insist on the correct pronunciation of "déjà vu" all that people will understand is that you think they're dumb for not speaking French.

Words are malleable, they can be played, they can be abused. Languages evolve and adapt. Writers make a living from recreating language over and over again and we admire the novelty. But the very reason that makes language socially useful makes it so unsuitable for science. It's imprecise, open to interpretation, dependent on a large number of unrecorded factors. As they say, if a thousand people read a book, they read a thousand different stories.

Mathematics on the other hand does not suffer from these shortcomings. Mathematical structures are defined to have certain properties. They're not open to interpretation and don't depend on the context. Mathematics is therefore highly reproducible and precise. If a thousand people read Einstein's field equations, they read the same equations.

There is nothing unreasonable about the efficiency of mathematics as a language because we call it efficient exactly because it is reproducible and precise. The actual surprise, as Wigner discusses in his essay, is that we find ourselves in an environment that has allowed us to discover many laws which lend themselves to a mathematical formulation. "The laws of nature," Wigner writes, "must already be formulated in mathematical language." Really? Pragmatic raises an eyebrow. What we actually know is that some laws of nature can be formulated in mathematical language. But does this mean our observations must be describable as mathematical law?

Pragmatic Physicists thinks of his ancestor, Pragmatic Neanderthal. She climbed to the top of the fire mountain and scanned the land to her feet. "Why is the human eye so effective at perceiving the all-there-is?" she mumbled to herself. "It is not," said the time-traveler who just appeared behind her, "It is due to natural selection that the surviving species are likely to be good at perceiving what is relevant to their survival. You do not in fact perceive most of the all-there-is." He was in the middle of explaining galactic rotation curves when she hit him over the head and sacrificed him to the Fire God. How effective is your math at describing her behavior?

Wigner carefully wrote about the efficiency *in the natural sciences*. But what, fundamentally, is not part of nature? If we draw the line between the natural sciences and the 'soft' sciences based on the extent to which they use math, then it is tautologically correct that mathematics is effective in the natural sciences. Nothing can be learned about the general efficiency of math if we only look at that what we describe with it, so let us look then at that what we do not describe by mathematics.

We do not, for example, formulate studies in the history of, say, 16th century Hip-Hop culture in mathematical quantors. It would be highly impractical, for not to say inefficient. Practical Historian, the great-aunt of Practical Physicist shudders at the thought! Aside from statistical analyses, history and sociology is still mostly verbally interpreted because it is extremely hard to quantify all the different nuances and aspects relevant to human culture.

If reductionism is correct and all of human culture is a consequence of the fundamental interactions among elementary particles, there is still an enormous amount of factors that would have to be taken into account, and most of them are unknown to us. And so we do not presently know whether it is possible at all to construct a mathematical model of human behavior that is any better than just watching the real system to see what happens. Practical Physicists reminds you that the point of doing science was not finding just any model, but a model that is useful.

It is certainly possible in certain cases to find useful mathematical models for human behavior, so this is not to say that one should not even try to describe sociology or history by math. Many examples exist where regularities can be found and applied, such as the motion of crowds [6] or the growth of cities [7]. But at least for now this is the exception rather than the rule.

Physicists like to think that it is possible in principle to develop a mathematical model for human behavior, and Pragmatic is no exception. But for such a model "in principle" to be of any scientific value "in practice" it must be more useful than an identical copy, and Pragmatic isn't sure at all that a mathematical model of human behavior could ever predict anything. At present, even the best computers on Earth cannot predict the folding of single proteins, something that happens in nature within milliseconds.

Practical physicist has been around for long enough to know that questioning Wigner's supposed "unreasonable effectiveness of mathematics" is blasphemy for his colleagues. "But, but," the readers with a PhD mumble and bounce on their chairs thinking Practicial Physicist would make good fodder for the Fire God. So let me give this to you, dear reader. Practical Physicist has no idea why we find ourselves in an environment that is configured so that it displays many recurring, reproducible, and often self-similar subsystems that can be greatly simplified thanks to properties such as cluster decomposition, scale separation, and locality. All he is saying is given that these subsystems do exist, it isn't surprising that math is efficient at describing them because the essence of mathematics' success is the delivery of simplified universal models that are reproducible and reusable.

This brings Practical Physicist back to the question whether all is describable by mathematics.

Models. Behaving. Gladly

Reminded of the role of mathematics in physics, Practical Physicist sees a way to move on. The little arrow that is science in his diagrams is not just an arrow, it is something he does constantly, something humans have done for thousands of years. He will look at examples to see which case fits.

So he thinks of himself as student calculating the energy levels of the hydrogen atom. He thinks of Einstein calculating the deflection of light on the Sun. He thinks of the Millennium Simulation [8] numerically recreating the formation of large scale structures. He thinks of the bump in a plot that we came to refer to as the Higgs boson. He thinks of the recent Nobel Prizes, of graphene, fibre optics, and supernovaes. He thinks of dark matter, superconductors, and quantum computing. And it dooms on him that none of the cases he depicted describe how science actually works.

Practical Physicist slaps his forehead and calls himself a fool because he took for granted what he learned, that we use models made of something called mathematic that are distinct from observations. But this has never has been so. Our models are observations, like that what they describe.

Think of any scientific explanation of any observable phenomenon. Take, say, the comet on its orbit around the sun. The process of science is not, as Practical Physicist had been taught, finding and using a map from mathematics to observation. No, the models that we use are subsystems of our universe, just like the system we want to describe (Fig. 2).

The model is you doing a calculation with pen on paper. It's a computer doing the calculation for you. It's a computer running a Monte Carlo algorithm. It's your

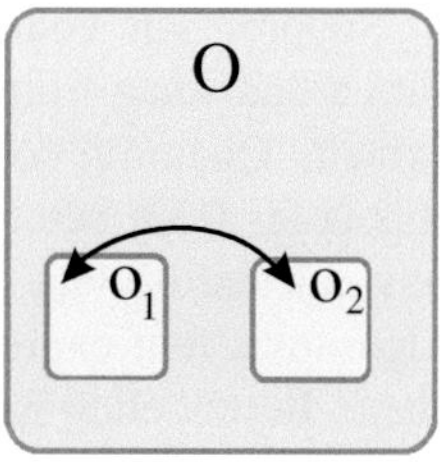

Fig. 2 Science works by mapping subsytems to each other.

student plotting a graph. In each of these cases we map one observation to another. We take the results of a calculation—a plot, a table, a number—and match them to another observation. If you think of Friedmann's equations and their solutions, you *are* a model. Maybe not a good one, all right, but the processes in your brain do exactly what we asked a useful scientific model to do.

That complex phase of the wave-function, the prime example for a mathematical structure that is supposedly unobservable? You have seen it dozens of times. It is perfectly possible to visualize complex numbers so that humans can perceive them. That it does no longer correspond to an observable after mapping it to another subsystem is irrelevant. Nobody ever demanded that to be so. All we want of the theory (the model including the map) is to reproduce observations, all we want is for it to be useful.

How to Do Science Without Mathematics

Practical Physicist is really excited now because suddenly he sees how to take mathematics out of the scientific process. To be a useful model, one that makes predictions or describes many observations, the model must not necessarily be described in the language of mathematics. All we need is to know how to construct subsystems that can be mapped to other subsystems in a process that is reproducible.

If Pragmatic Neanderthal had not heartlessly sacrificed the time-traveler to the Fire God, the time-traveller might have given her a little black box with the amazing property of being able to predict volcano eruptions a month in advance. Volcanologists today would certainly like to get their hands on that little black box. But for Pragmatist, making a prediction with this box would not be science because it is not reproducible. If the box breaks, we'd be back to start.

But consider the time-traveler told you how to construct the little black box. Now you could make one yourself. You would get to know that what it is made of and you would develop an understanding for it. You could experiment with it, learn how to modify it, and how to use it for other tasks. For all practical purposes the little black boxes could be just as good as mathematical models. They could be even better because they could work in cases when mathematics does not work, if only we could give up the idea that all models have to be formulated in mathematical language.

Suppose you could initialize an adiabatic quantum computer so that it evolves into a final state from which you can read off the result of the folding of a certain protein. Quantum computers that can be constructed right now are so simple that this is far from being possible, and moreover they have never been put to use on any calculation that could not have been done also by a normal computer, using algorithms that express mathematical manipulations. But the future may be right there: In modeling nature directly, leaving out mathematics as the middle man, by directly comparing different types of subsystems.

It is not only in quantum computing that we can sense a beginning of this development. It is also implicit in the discovery and use of dualities in physics [9] and in systems that are "analogue" to each other, both of which is just a way of saying that certain observations can be mapped to each other. These dualities have been found analyzing mathematical structures, yes, but may it be that they hold beyond that?

Take as an example Analogue Gravity [10, 11]. Analogue Gravity exploits the fact that perturbations in a fluid propagate like particles would in certain gravitational fields. The best known example is Unruh's dumb hole—a fluid in flow that creates a 'sonic horizon' for perturbations on its surface if the speed of the fluid's flow exceeds the speed by which perturbations can travel. This sounds like a curious coincidence, but it can be shown that such gravitational analogues exist in a large variety of cases. Take an imaginative leap now and suppose you could model supergalactic structure formation by observing the behavior of a suitably chosen condensed matter system in the laboratory. It would be much like running a computer simulation, just that you don't write the algorithm. You don't even need to know if one exists.

A similar statement can be made about the AdS/CFT duality [12–14]. Physicists presently use this duality as a calculation tool, but imagine you would instead create one system in the laboratory and use it to extract predictions about another system. You would not be doing a mathematical calculation, you would just map one subsystem to another.

You might argue now that the difference is one of understanding, that without understanding the use of the model is limited. Maybe that is so, but note that your idea of understanding is greatly biased by your experience that understanding can include a mathematical framework. But if a model is the only model for a specific purpose, any model is useful.

In his essay "The End of Theory" [15] Chris Anderson argued that the end of theory is nigh because computers are able to process ever increasing amounts of data and infer regularities from this data without ever developing a human-readable theory. Anderson wasn't concerned with the power of mathematics, but critics claimed that this would stall scientific progress because no true understanding is achieved, a point also relevant for the idea of science without mathematics. So let us reflect on this.

Anderson is both right and wrong. He is of course exaggerating when he says theory is coming to an end because we still need someone to write the computer code. But he is right that the relevance of skillful simplification decreases with larger computing power, and that it is ultimately unnecessary for humans to be able to formulate a simplified underlying mathematical law to do science, as long as the results are useful.

Mathematics is certainly not coming to an end. Mathematical models are and will probably remain the best way to understand observations. Pragmatic scientist does not want you to give up on math. He is just pointing out a way to still create scientific models in cases mathematics does not work.

Even if you do not believe that there are any observations for which we cannot find a mathematical model, a non-mathematical model that works by comparing observations still can be a step on the way to finding a mathematical model. You might for example have succeeded in describing the dynamics of certain fluids by means of gravity had you exploited the observed similar behavior of these systems. It's just that we had already been using the hydrodynamical equations before we discovered the similarity to gravity.

Finally, let us come back to the question whether M contains itself or doesn't. It's the classical example of a question that cannot be answered within the mathematical framework that we are using. But we already knew from the very beginning that using math to classify what a language beyond math can do would not be self-contained. Instead, think of it as a bootstrap approach to beyond-the-math-physics. The classification in Veng diagrams is a first step, but it cannot provide all answers that we would like to have. Likewise, different types of logic in which the depicted cases are not mutually exclusive constitute a generalization of this approach. Pragmatic's observation that science can proceed without necessarily using mathematics is a way to shed light on which of the philosophies of Fig. 1 is most compatible with nature. We may never be able to settle this question entirely, but we may be able to learn if there are limits to mathematics, where they are, and how close we can get to mathematical perfection.

Conclusions

Pragmatic Physicist hopes that you read his whole essay, but he is too pragmatic to think your reading habits are any better than his. So in case you jumped here from the abstract, a brief summary:

Scientific models must be useful to describe our observations, either by making predictions or by explaining many observations with one model. There is however no particular reason why models must be mathematical to be useful. Mathematical models today work not by identifying abstract mathematical structures with observations, but by identifying observable realizations of mathematical structures with observations. Scientific models, in general, identify different observations with each other. There is then no specific reason why this identification must necessarily involve mathematics. One could do science by directly trying to identify the observables belonging to different subsystems, and in this way learn about the effectiveness and limits of mathematics.

If technology continues to progress and flourish, the day will come when we can link human brains and language will become an unnecessary intermediary of communication. In the very same way, mathematics may one day become an unnecessary intermediary of science.

References

1. Smolin, L., "The Trouble with Physics", Houghton Mifflin Harcourt, New York (2006).
2. Tegmark, M., "*The Mathematical Universe*," Found. Phys. **38**, 101 (2008) [arxiv:0704.0646 [gr-qc]].
3. Hossenfelder, S., "The Mathematical Universe", Backreaction, Oct 11 2007, retrieved Jan 4 2015 at http://backreaction.blogspot.com/2007/10/mathematical-universe.html.
4. Twain, M., "The Awful German Language," in "A Tramp Abroad," American Publishing Company, Hartford, Connecticut (1880), retreived Jan 4 at http://www.kombu.de/twain-2.htm.
5. Wigner, E., "*The Unreasonable Effectiveness of Mathematics in the Natural Sciences*," Communications in Pure and Applied Mathematics, vol. 13, No. I (February 1960).
6. Hughes, R. L., "The Flow of Human Crowds," Annual Review of Fluid Mechanics **35**, 169–182 (2003).
7. Bettencourt, L., and West, G., "*A unified theory of urban living*," Nature **467**, 912913 (21 October 2010).
8. Lemson, G., [VIRGO Collaboration], "Halo and Galaxy Formation Histories from the Millennium Simulation: Public release of a VO-oriented and SQL-queryable database for studying the evolution of galaxies in the Lambda-CDM cosmogony," arXiv:astro-ph/0608019.
9. Polchinski, J., "*Dualities of Fields and Strings,*" arXiv:1412.5704 [hep-th].
10. Unruh, W. G., "Experimental black hole evaporation," Phys. Rev. Lett. **46**, 1351 (1981).
11. Barcelo, C., Liberati, S. and Visser, M., "Analogue gravity," Living Rev. Rel. **8**, 12 (2005) [Living Rev. Rel. **14**, 3 (2011)] [arXiv:gr-qc/0505065].
12. Maldacena, J. M., "The Large N limit of superconformal field theories and supergravity," Adv. Theor. Math. Phys. **2**, 231 (1998) [arXiv:hep-th/9711200].
13. Witten, E., "Anti-de Sitter space and holography," Adv. Theor. Math. Phys. **2**, 253 (1998) [arXiv:hep-th/9802150].
14. Gubser, S. S., Klebanov, I. R., and Polyakov, A. M., "A Semiclassical limit of the gauge / string correspondence," Nucl. Phys. B **636**, 99 (2002) [arXiv:hep-th/0204051].
15. Anderson, C., "The End of Theory," Wired Magazine June 23 2008, retreived Jan 5 at http://archive.wired.com/science/discoveries/magazine/16-07/pb_theory.

The Descent of Math

Sara Imari Walker

Abstract A perplexing problem in understanding physical reality is why the universe seems comprehensible, and correspondingly why there should exist physical systems capable of comprehending it. In this essay I explore the possibility that rather than being an odd coincidence arising due to our strange position as passive (and even more strangely, conscious) observers in the cosmos, these two problems might be related and could be explainable in terms of fundamental physics. The perspective presented suggests a potential unified framework where, when taken together, comprehenders and comprehensibility are part of *causal structure* of physical reality, which is considered as a causal graph (network) connecting states that are physically possible. I argue that in some local regions, the most probable states are those that include physical systems which contain information encodings—such as mathematics, language and art—because these are the most highly connected to other possible states in this causal graph. Such physical systems include life and—of particular interest for the discussion of the place of math in physical reality—comprehenders able to make mathematical sense of the world. Within this framework, the descent of math is an undirected outcome of the evolution of the universe, which will tend toward states that are increasingly connected to other possible states of the universe, a process greatly facilitated if some physical systems know the rules of the game. I therefore conclude that our ability to use mathematics to describe, and more importantly *manipulate*, the natural world may not be an anomaly or trick, but instead could provide clues to the underlying causal structure of physical reality.

Anthropic arguments are often used to explain away some of the most perplexing questions of our existence. Among these is the question of why the values of the constants of nature seem *surprisingly* well suited for life—indicative of a problematic degree of fine-tuning (dubbed the "fine-tuning problem"). However, one can reason, based on anthropic arguments, that the constants must be such as they are; otherwise we would not be here to ask about them [1]. We can similarly employ anthropic

S.I. Walker (✉)
School of Earth and Space Exploration and Beyond Center for Fundamental Concepts in Science, Arizona State University, Tempe, AZ, USA
e-mail: sara.i.walker@asu.edu

A. Aguirre et al. (eds.), *Trick or Truth?*, The Frontiers Collection,
DOI 10.1007/978-3-319-27495-9_16

arguments to "solve" the problem of the rapidity of life's emergence, which happened almost as soon as conditions were favorable, and the related problem of an apparent arrow of increasingly biological complexity with time; both are necessary for us to be here and now asking how it could have happened [2]. While such appeals to anthropocentrism may not be entirely satisfactory to some, the reasoning is logically sound, challenging whether further explanation beyond anthropic arguments is indeed necessary.

There is however at least one anthropic feature of the universe as we observe it that cannot be adequately dismissed with these kinds of arguments. Namely, *why does the universe seem comprehensible and why should beings like us be here to comprehend it*? There appears to be no necessary reason why the universe should make sense to a subsystem of itself, and no necessary reason that we humans, being such a subsystem, should be able to make sense of it. Certainly we could exist without comprehending the world in a deep mathematical way (or for that matter making any sense of it at all (and indeed, one might argue that this is the *modus operandi* for most of us!)). While we would not be having an intellectual debate on the issue if we were not comprehenders, there is nothing about our existence that depends on our ability to make mathematical sense of the world in the same way that our existence is dependent on the fine-tuned value of the fine-structure constant (which if changed by just 4 % would render formation of Carbon via stellar fusion impossible [3]). The situation is stranger still if we consider not only that we can make sense of the world by utilizing descriptive tools such as math, but that we can learn to do so as individuals, roughly on the timescale of a university education (see e.g., [4] for discussion on this intriguing point).

It is easy enough to ignore these kinds of oddities, either as bizarre flukes or as pointless questions from the scientific standpoint. However, reality draws no dividing line between "artificial" and "natural" or, more to the point, "comprehender" and "everything else", as we have a tendency to do from our anthropocentric vantage point. Thus, the observational facts that the universe seems to make some sense to us, and that we can indeed (at least in part) make sense of it, must be taken into consideration in scientific discourse if we want to fully understand the universe in which we actually live (e.g., one that includes comprehenders).

In example, a perfectly good scientific account exists for how Earth and other planetary bodies and natural satellites were formed via accretion from the solar nebulae, some 4.5 billion years ago. We do not have a similar explanatory framework that encompasses the "anti-accretion" happening in the Earth system today, where artificial satellites are being launched into orbit at an increasing pace [5]. The most important difference is that "anti-accretion" requires comprehenders—specifically, the existence of physical systems with *knowledge* of Newton's laws. To state this distinction more explicitly, the problem here does not arise in our descriptions of the orbital mechanics of the International Space Station (ISS) versus that of the Moon. Both would be much the same, with minor variation (e.g., could be explained using Newtonian gravity, or more precisely general relativity). If instead, however, you were to ask how each was *caused* to enter its orbit, in the case of the Moon you could provide a perfectly good account based on the initial conditions of the Solar System and the laws of physics. In the case of the ISS, that account would necessarily also include *physical systems with knowledge of the laws of gravitation.*

Making this distinction about causation may seem unnecessary, but it leads to a significantly different perspective on the structure of reality in the local vicinity of comprehenders, which I explicate in this essay. This perspective permits the incorporation of the existence of comprehensibility and comprehenders in a unified framework where, when taken together, comprehenders and comprehensibility are integral to the *causal structure* of physical reality. As such, rather than being an anomaly or bizarre fluke, the fact that we use information encodings—such as mathematics, language and even art—to describe and more importantly *manipulate* the world may in fact be a highly probable state in the space of all possible states of physical reality. It therefore could be explainable in terms of fundamental physics. In this formulation, the "*descent of math*" is an undirected product of the evolution of physical reality (under certain assumptions), just as the "*descent of man*" is an undirected product of the process biological evolution [6] (that is, neither requires design, although both might give the appearance of design (see e.g., [7] for discussion on this matter)).

Life and Physics: Two Roads Diverged in a Wood

In physics we are trained to think in terms of initial conditions and deterministic, fixed laws of motion (the "prevailing conception" as discussed by Deutsch [8]). This has been an incredibly powerful approach to understanding systems as diverse as the interior of stars, superconductivity and swinging pendulums. However, at least thus far, this approach has fallen short of providing an adequate explanatory framework for life or mind. A challenge is that in describing life and related processes, we often use words such as "signaling", "symbols" and "codes"—that is, biology is cast in the language of *information*. The very concept of information, however, at present is not readily reconciled with a narrative cast solely in terms of initial conditions and fixed deterministic laws. A simple way to conceptualize this rift is to recognize that in order for something to carry information, it needs to have at least two possible states, e.g., to say something carries 1 bit of information means that it could have been in one of two possible states. Fixed deterministic laws only allow one outcome at a given time t for a given initial state at $t = 0$, and therefore do not allow the possibility of more than one possibility. That is, they do not allow the possibility of *counterfactuals*—roughly described as situations that do not happen, but could. Our current laws of physics permit a host of unrealized universes that could happen, but don't. Starting from a given initial condition, the future state of the universe as described by fixed dynamical laws is set for all time on one—*and only one*—of what could be many possible trajectories through state space.

We should contrast this with what we know of biology, which appears to be incredibly path-dependent. A good example is the path-dependence of biological evolution. Starting from the same initial state, biological systems trace out an enormous array of alternative trajectories through the process of evolution. Thus, a common statement in evolutionary biology such as "we share a common ancestor with X", where X is a chimpanzee, fungi, or archaea, effectively is meant to tell us how far in the past we once shared a common initial state (to rough approximation, a genome) with species X. Each separate species evolved its unique features due to the peculiar and specific selection pressures it has seen through its evolutionary history, both due to the environment and competition with other organisms. Thus, in biology we are well acquainted with the fact that the current state is a function of (evolutionary) history. It is difficult, if not impossible, to write an equation of motion for such historical processes. The challenge is the "state-dependent" or "self-referential" nature of this kind of dynamical system [9]: the manner in which biological systems evolve through time is a function of their current state, such that the dynamical rules themselves change as a function of the current state [10]. The "laws" of biology therefore appear to be time dependent (this is a hallmark of self-referential systems and life, see e.g. [10, 12] for discussion). This naturally leads to historically dependent trajectories: the current state will depend on the sequence of previous states [11], masking dependence on initial conditions. State-dependent dynamics are deeply tied to the role of information in life: in part, the information encoded in the current state dictates what the system will do next.

What is intriguing about this situation is that biology appears to be taking advantage of precisely what is lacking in our ability to unify physics and information: biological systems are capable of taking many possible paths through state space, which must somehow be distinguished by their use of information. This observation has led many to suggest that the physics of biology *is* the physics of information, and in particular for my collaborator Paul Davies and I to suggest that the origin of life itself coincides with the emergence of physical systems where information plays a causal role (see e.g. [12] and references therein) perhaps representing an entirely new frontier in physics. Cast in the language of the discussion presented here, biological systems appear to represent a different kind of physics, one where information, in an as yet unspecified sense, seems necessary to define the trajectories taken through state space. To connect this story to the role of comprehenders, and our use of math to comprehend reality, we will need to take a detour and explore the structure of what is possible under the known laws of physics.

You Can't Get There from Here

Our world seems inordinately complex, being chock full of interesting things like bacteria, ant colonies, humans and cities. With our current formulation of physics, *all* of this complexity must be explained in terms of initial conditions and *fixed* dynamical laws. This leads to an odd accident as far as the status of initial conditions is concerned in the way we do physics. Since we do not seem to have the freedom to play with the laws, the forgoing indicates that any explanation we have for why the world is such as it is, and not any other way it could be, must be included in the specification of the initial state.

Take for example, a state of reality where Germany defeats Argentina to win the World Cup. Certainly an alternative state of the world—a counterfactual—might be Argentina defeating Germany. Neither scenario, describing Germany or Argentina emerging as the victor, violates the laws of physics, so it cannot be the laws themselves that distinguish which event happens.[1] This leaves open only the option that the initial condition is the distinguishing factor, i.e., that the universe started in a very special initial state in which it was encoded that the Germany would win the World Cup on a planet called Earth, orbiting a humdrum star, in AD 2014. In the framework of initial conditions and fixed laws of motion as the sole descriptor of reality, the more complex the world becomes, the more special among the set of possible initial states must be the one that actually produced it (i.e., the more fine-tuned it becomes). Apart from feeling unsatisfied with the idea that the universe's future for all time is laid down in its initial state (which indeed must necessarily be very special to explain our world), the focus we have in physics on initial states and fixed laws of motion is also somewhat at odds with a general feeling some of us may share that anything that is possible (allowed by the laws of physics) should be able to be caused to occur.

In statistical mechanics we base many of our calculations on the assumption of metric transitivity, which asserts that a system's trajectory through phase space will eventually explore the entirety of its state space—thus everything that is physically possible will eventually happen. That is, anything that can be caused to occur by the laws themselves will happen, if only we wait long enough. While there are many examples of isolated systems that do not obey metric transitivity (possibly as a result of their underlying causal structure), the feature that is important for discussion here is that true metric transitivity makes initial conditions irrelevant, since every state will eventually be visited. In metrically transitive systems, "special" initial conditions that lead to restricted orbits through phase space, isolated from other possible trajectories, are excluded (are of measure zero).

A key point is that if we require specialness in our initial state (such that we observe the current state of the world and not any other state) metric transitivity is violated. It is then not necessarily possible to get to any other physically possible state, even those that may be equally consistent and permissible under the laws of physics. This leaves us in a bit of a perplexing situation, as we require special initial conditions to explain the complexity of the world, but also have a sense that we should not be on a particularly special trajectory to get here (or anywhere else) as it would be a sign of fine-tuning of the initial conditions. More simply put, a potential problem with the way we currently formulate physics is that you can't necessarily get everywhere from anywhere.

[1]This may be a point of contention for some, as one might argue that Germany won due to physical superiority, but that is a different sort of physical argument than the one being put forth here.

Physical Reality and the Art of the Possible

The key point of the above discussion is best clarified if we distinguish "possible" states of the world, defined as those that are not forbidden by the laws of physics, and "physically accessible" states of the world, which are those that are achievable from a given state (either in its "past" or "future"). Metric transitivity requires that all possible states are physically accessible, that is we assume systems where

$$\mathbb{N}_A \approx \mathbb{N}_P$$

where $\mathbb{N}_P$ are the number of physically possible states of the world, defined as those allowed by the laws of physics, and $\mathbb{N}_A$ are those that can be realized (are physically accessible) from a given starting point.

However, a physics where $\mathbb{N}_A \approx \mathbb{N}_P$ has very strict constraints on its causal and informational structure and is not necessarily always metrically transitive. It is simplest to envision what these constraints are by considering *a causal graph (network) for reality*, where two instantaneous states of the universe are connected if, under the laws of physics, one state maps to the other. Edges connect nodes with their possible causes (states that map to the given node under the laws of physics) and effects (states that a given node maps to under the laws of physics). The constraint $\mathbb{N}_A \approx \mathbb{N}_P$ imposes a very restricted topology on this graph. For one, each possible state must have only one cause and one effect,[2] and therefore each node has at most two edges. This constraint also means that microscopic reversibility holds and that no information is lost in the mapping between states (this is true because a cause or an effect are fully specified by looking at the current state, since the mapping is one-to-one). In such systems, the trajectories through state space follow fixed paths along closed causal loops. In many dynamical systems, such as cellular automata, these causal loops are disconnected, so while $\mathbb{N}_A \sim \mathbb{N}_P$, metric transitivity does not hold because there are isolated parts of the graph. This is probably true in the real world, but we often do not consider the underlying causal structure in our theories, particularly in statistical physics where it is often assumed that any microstate can transition to any other with equal probability.

By contrast situations where $\mathbb{N}_A \not\approx \mathbb{N}_P$ arise when the causal structure is such that multiple causes map to one effect (or a cause has multiple effects). Nodes can have any number of edges (up to a number connecting a node to any possible state). An important point about this causal architecture for physical reality is that information is lost in mapping between states for a many-to-one map: if you ran the system in reverse, there would be uncertainty in which state was the initial cause. This has potentially deep connections to information loss and the emergence of an arrow of time, which is normally associated with information loss due to coarse-graining [13].

[2]This constraint imposes all possible states of the world to have a one-to-one deterministic mapping connecting each state to at most two other states per time step.

In this formalism, states are connected when they share information.[3] Importantly, the number of states accessible from a node (its effects) is precisely the number of its out-directed edges in the network graph of reality. Likewise, the number of states a given node can be accessed from (its causes) is number of in-directed edges.

An example of why any of this should matter may be in order, so we will return to our initial discussion of Earth and its satellites. Consider the set of all states of reality that correspond to Earth plus satellites. This set may be partitioned into two subsets: states where some physical systems have knowledge of Newton's laws[4] (i.e., comprehenders, such as humans) and states where there is no such knowledge. One example of the latter is Earth with one natural satellite—the Moon, which was the state of Earth for its 4.5 billion year history prior to 1957 when Sputnik was launched. It is physically possible that Earth might have also had no Moon, two captured asteroids for Moons (as is the case for Mars) or a potential host of alternative smallish rocky bodies orbiting it. All are equally viable states of the world consistent with known physics. It is also physically possible for the Earth to have any number (within resource constraints) of artificial satellites or space junk. However, the latter set of "anti-accreting" states, while possible, is not accessible (by this I mean *not* encoded in the initial conditions) in the absence of comprehenders and the technology they create. Thus, when comparing the size of the state space of Earth plus satellites, the state space is *much larger* if artificial satellites and the comprehenders that launch them to space are included, i.e., if their exist physical systems that contain knowledge of the laws of gravitation (as well as good engineers).

The above suggest there may in fact be another kind of edge in the causal graph of reality in addition to those set by the laws of physics—they are those set by physical systems that contain knowledge of the laws of physics. In an "*explanatory graph of reality*", some of the causal edges between states are set by the existence of physical systems that contain knowledge about other possible states. Here, 'knowledge' need not be anything nearly as sophisticated as the law of universal gravitation, but can include simpler rules encoded in physical systems, such as that swimming up a chemical gradient will likely result in finding a new food source, as is the case for bacterial chemotaxis. *Knowledge* in the sense presented here is therefore directly proportional to the number of counterfactuals about *what could be caused to happen*, instantiated in a given physical system. This approach therefore has some promise for quantifying the murky concept of "knowledge" in physics, precisely identifying the knowledge in a physical system with the number of out-directed edges in the explanatory graph. Physical instantiation of knowledge like Newton's laws is particularly powerful since it connects a large number of states that would otherwise be disconnected (hence it is corresponds to a large number of "explanatory" edges or knowledge and is deemed a "law").

[3]This last point is most obvious for the case of one-to-one mapping where no information about past states is lost in the mapping.

[4]To be perfectly accurate I should include general relativity in this argument, but the argument stands the same regardless of what our current theories to describe reality are, so long as they permit new states of the world to be realized that could not be realized in the absence of such knowledge.

There is a second important reason why knowledge of Newton's law manifests differently in the explanatory graph than the edges set by Newton's law itself. To illustrate this, we return again to our two aforementioned examples of states of the Earth system with natural and artificial satellites. In the case of the Earth and Moon, described under a scenario of initial conditions and deterministic laws, *only* those states pre-ceding the giant Moon forming impact could have "caused" the Moon to form. A second ISS, by contrast, could easily be launched to space within a few months if there was sufficient will power to do so. Comprehenders therefore have the property that they can more reliably cause a transformation to occur (in the sense that it can happen again) than the laws of physics alone can. This notion of reliability is an important concept in constructor theory [7, 8], where "constructors" are identified with reliable causes.

Reliability implies that states that arise as a result of traversing a trajectory connected by knowledge should, on average, retain the knowledge instantiated in past states along the trajectory (such that transformations that were possible in the past remain possible). Thus, connectivity should be "heritable" among states instantiating knowledge. Such *heritable connectivity* could provide a physical explanation for the ubiquity of reproduction in biological systems, reproduction is a surefire method of ensuring heritability of information encodings necessary to instantiate knowledge (retain causal edges).

The combination of increased connectivity and reliability in the vicinity of comprehenders or knowledge-creating systems provides a potential explanation for the situation we find ourselves in as comprehenders occupying a seemingly comprehensible reality. When considering the number of states of the world consisting of Earth plus satellites, very many more states are reliably accessed from states in which some physical systems have knowledge of the laws of gravitation than from those where there is no such information. Thus, in a network view of reality, where causal edges connect all physically possible states, nodes with physical systems (comprehenders) that contain knowledge of Newton's law are more highly connected to other nodes within the space of all configurations of Earth plus satellites.

The essential point of this argument in connecting the existence of comprehenders and comprehensibility is that states of reality containing physical systems with knowledge (comprehenders) should be probable based on the connectivity of state space, once one accounts for the underlying causal structure. What I mean by probable does not strictly refer to counting the frequency of a state (as is traditionally done in statistical physics), which does not account for the existence of any underlying causal structure. The most probable states here are those that are the most highly connected via knowledge (have the largest number of edges) in the causal graph because they are the most likely to be visited, if one assumes a random walk along the graph with the principle of reliable causes. I argue that those most probable states should include comprehenders, because through knowledge, comprehenders connect many states of reality that would otherwise be unconnected and do so more reliably than the laws of physics along (connectivity is heritable). These states must also be comprehensible, as otherwise knowledge would not result in connections to those states. Both comprehenders and comprehensibility are therefore required to create causal edges.

Life, Information and ... Math!

We should not necessarily immediately assume that reality is structured such that all states are accessible, or reliably accessible (and in fact as noted above, for reversible dynamical systems with $\mathbb{N}_A \approx \mathbb{N}_P$ all states are accessible from somewhere in the causal graph, but in most cases I think we would find none are from everywhere). *What kind of scenario then would allow the most states within a local region of the causal graph of reality to be realizable*? I suggest this requires physical systems that actively use information to move through state space; in short it requires life, mind and related processes. The impact on the world when we discover a new physical law—e.g., a pattern in how the world works—behaves much more like the process of biological evolution, than (perhaps ironically) it does like the physical systems we describe with those laws. Biological evolution accesses many states of the world from similar starting conditions. Likewise, the discovery of new laws of nature allows many states of the world to potentially be accessed that were not accessible before their discovery. At their core, both of these processes have one important thing in common: multiple states of the world can be caused to occur, which are not explicitly encoded in the initial condition (both necessitate states which are highly connected).

The physics necessary requires more than just an initial condition and deterministic law—*it requires information*. Physical systems, which encode information about other states of the world (connected by an edge) and where that information in part defines their trajectory through state space, will traverse non-trivial dynamical trajectories that are history and path-dependent. Although I have been discussing this dynamic as though time is discrete, discrete time is not central to the argument.

In the framework presented, the most probable among all possible states of physical reality are those that include physical systems which contain information encodings—such as mathematics, language and art—because these are the most highly connected states in the local state space of what is possible. That is to say, they have the highest number of reliable edges in the network of the possible states of the world. A random walk on the network of all that is possible will spend the most time on highly connected nodes, and is most likely to follow a trajectory where the connectivity of nodes *increases* as a function of time. This suggests a generalization of the second law of thermodynamics to the scenario where $\mathbb{N}_A \not\approx \mathbb{N}_P$, where instead of the most commonly visited states being those which are most probable, it is instead those *that are most highly connected*. This formalism naturally accommodates an arrow of increasing knowledge, corresponding to an arrow pointing towards states that have increasing information about other possible states of the world [12]. Such an arrow of knowledge would likely have many of the hallmarks associated with the apparent arrow of increasingly complexity observed in our biosphere. Within this framework, the descent of math is an undirected, but not unexpected, outcome of the evolution of the universe, which will tend toward states that are increasingly connected to other possible states of the universe, a process greatly facilitated if some physical systems know the rules of the game. This framework is still under development, but it shows promise as both *comprehenders*—defined as physical systems where instantiated information about the world (knowledge) in part determines their trajectory through state space, *and comprehensibility*—defined in terms of states that can share information (edges created by knowledge), are naturally accommodated. Our ability to use mathematics to describe, and more importantly *manipulate*, the natural world may not be an anomaly or "trick", but instead could provide clues to the causal structure of physical reality.

Acknowledgments The author wishes to thank Paul Davies and Chiara Marletto for comments, and in particular to thank Chiara Marletto for the terminology of "explanatory graph" to describe the network of reality containing comprehenders and the insight that heritable connectivity is essential to its formulation. Additional thanks to the FQXi community for the lively discussion and insightful comments on this essay

References

1. Barrow, J.D.; Tipler, F.J. (1988) The Anthropic Cosmological Principal. Oxford University Press.
2. Davies, P.C.W. (2008) The Goldilocks Enigma. Mariner Books.
3. Barrow, J.D. (2001) Cosmology, Life and the Anthropic Principle. *Ann. New York Acad. Sci.* **950**, 129–153.
4. Davies, P.C.W. (1990) Why is the Physical World so Comprehensible? *Complexity, Entropy, and the Physics of Information. SFI Studies in the Science of Complexity.* vol. VIII, Ed. W.H. Zurek, Addison-Wesley Press.
5. The term "anti-accretion" was coined by David Grinspoon.
6. Darwin, C. (1871) *The Descent of Man, and Selection in Relation to Sex.* John Murray Press
7. Marletto, C. (2015) Constructor theory of life. *J. Roy. Soc. Interface*, **12**, 20141226.
8. Deutsch, D. (2013) Constructor Theory. *Synthese*, 190, 4331–4359.
9. Hofstader, D. (1999) *Godel, Escher, Bach: An Eternal Golden Braid.* Basic Books.
10. Goldenfeld, N. and Woese, C. (2011) Life is physics: Evolution as a collective phenomenon far from equilibrium. *Ann. Rev. Cond. Matt. Phys.***2**, 375–399.
11. Pavlic, T., Adams, A., Davies, P.C.W. and Walker, S.I. (2014) Self-referencing cellular automata: A model of the evolution of information control in biological systems. arxiv: 1405.4070.
12. Walker, S.I. and Davies, P.C.W. (2013) The algorithmic origins of life. *J. Roy. Soc. Interface,* **6**, 20120869.
13. Davies, P.C.W. (1977) *The Physics of Time Asymmetry.* University of California Press.

The Ultimate Tactics of Self-referential Systems

Christine C. Dantas

To the question which is older, day or night,
he [Thales of Miletus] replied: —
Night is the older by one day.
Thales of Miletus [624 BC–546 BC]

Either mathematics is too big for the human mind
or the human mind is more than a machine.
Kurt Gödel

Abstract Mathematics is usually regarded as a kind of language. The essential behavior of physical phenomena can be expressed by mathematical laws, providing descriptions and predictions. In the present essay I argue that, although mathematics can be seen, in a first approach, as a language, it goes beyond this concept. I conjecture that mathematics presents two extreme features, denoted here by *irreducibility* and *insaturation*, representing delimiters for self-referentiality. These features are then related to physical laws by realizing that nature is a self-referential system obeying bounds similar to those respected by mathematics. Self-referential systems can only be autonomous entities by a kind of metabolism that provides and sustains such an autonomy. A rational mind, able of consciousness, is a manifestation of the self-referentiality of the Universe. Hence mathematics is here proposed to go beyond language by actually representing the most fundamental existence condition for self-referentiality. This idea is synthesized in the form of a principle, namely, that *mathematics is the ultimate tactics of self-referential systems to mimic themselves.* That is, well beyond an effective language to express the physical world, mathematics uncovers a deep manifestation of the autonomous nature of the Universe, wherein the human brain is but an instance.

This essay received the 4th. Prize in the 2015 FQXi essay contest: "Trick or Truth: the Mysterious Connection Between Physics and Mathematics".

C.C. Dantas (✉)
Instituto de Aeronáutica e Espaço, AMR/IAE/DCTA, São José dos Campos, Brazil
e-mail: christineccd@iae.cta.br; ccordulad@gmail.com

A. Aguirre et al. (eds.), *Trick or Truth?*, The Frontiers Collection,
DOI 10.1007/978-3-319-27495-9_17

Introduction

Outline—The literature on the connection between mathematics and physics is vast, and will not be reviewed here (a good start is given by Penrose [1]). The present theme is about one of most profound and disturbing mysteries ever touched by the human mind, and it is obviously very hard to find a definite answer at our present state of knowledge. The best one can do about that gigantic question is to offer some perceptions, specially considering a short text such as this.

The outline for this essay is the following. I review the idea that mathematics is, essentially, some kind of language, and argue to what extent this position can be sustained as a fundamental one. Here, mathematics is seen as something more than a language. I introduce two conjectures for mathematics and identify how these conjectures apply to nature. Finally, I state how these conjectures serve as guide for a principle that amalgamates mathematics and physics from an unusual point of view.

To what extent is mathematics a language? — A language is a form of expression and communication, which can be realized by different means, but it also involves some kind of transformation between two domains. The study of the methods, development, and applications of mathematics, indicates that it operates as a language.

Internally, mathematics is based on some set of non-contradictory, fundamental notions (axioms), which are constructively enlarged into more sophisticated statements and relations (theorems), by the incremental use of logical operations. Externally, mathematics mediates the description of relations from the sensible domain of phenomena into the domain of logical associations, which are, by themselves, internally consistent.

Actually, for whatever purpose, mathematics seems to be "overly effective", because it can be used to describe, to a certain extent, anything one desires.[1] There is always a means, under varying degrees of limitations, in which any sensible object, process, or relation can be expressed by a formal, deductive system, such as mathematics. The ubiquity of mathematical representations and transformations is sometimes regarded a trivial "fact of the world", because it just implies that mathematics is wide and internally consistent enough to embrace any condition or aspect. But such a language must necessarily be of a very special kind, considering its "unreasonable effectiveness" [2] to describe physical phenomena. Is there a "natural selectiveness effect", based perhaps on the own rigorous nature of mathematics, able explain its universality of application? Why is nature so mathematical?

I believe that there is no "natural" explanation for the "unreasonable effectiveness"—or even the "overly effectiveness"—of mathematics if we regard it purely as a language, that is, from a fundamental point of view. My justification is based on the following observations:

[1] For occurrences of the Golden Ratio in the natural world, see, e.g., Ref. [6]. For a beautiful example of two different partial differential equations (PDEs) describing the same physical phenomenon, see Ref. [7]. A limiting procedure that suggests an explanation for the wide applicability and universality of some integrable PDEs can be found in [8].

(a) When mathematics is seen as a language, it is merely a "receptacle for logically configured relations" that summarize the essence of a physical phenomenon. Although it is, in a first approach, a reasonable view, it does not explain where the rules and meanings of the logical relations come from. The only possible way we could justify it as an "effective receptacle" would be in a *normative sense*, that is, in a prescriptive way. *But such a normative way does not offer any intrinsic or fundamental connection with the natural phenomena, except if imposed arbitrarily.*

(b) When a physical law is expressed mathematically, no evidence for the "truth" regarding the law actually surfaces, but only a guarantee for the logical foundation of the outcomes resulting from the implemented assumptions. In fact, any application of mathematics (seen as a form of language) to physical problems just brings forth a *relative valorization of their logical evidence, but not the evidence, per se, of their truth.*

I firmly believe that mathematics is more than a "transformation machine".[2] I regard mathematics as a language to the extent that it indeed *serves* as a language, but such a serviceability cannot be an admissible explanation for why nature is mathematical. In order to address this question, I raise two conjectures, which will serve as a basis for looking at the problem from a different perspective.

Two Conjectures for the Foundations of Mathematics

The proposed conjectures are philosophical, a fact that could be unattractive to some readers. However, these ideas could eventually be expressed in a more concrete or formal way, so they should be regarded as preliminary for the purposes of the present essay. It is clearly very hard[3] to develop an independent methodology to avoid the ironic situation of using mathematical principles themselves in order to explain mathematics. This is not, evidently, the purpose here, so we limit to qualitative statements, on a more "meta", abstract level.

I. IRREDUCIBILITY: Mathematics is Irreducible to Anything Else that is Not Itself Mathematically Expressible.

What it means—Mathematics cannot be primarily reduced to anything different from itself. Any attempt to characterize mathematics, methodically, will necessarily involve the use of mathematically defined constituents and procedures (representations), so that *an ultimate regression that is not mathematically based is impossible.*
Conjecture's nickname—The miniature extremeness of mathematics.

[2]Although very useful, mathematics would be ultimately limited to transform "emptiness into emptiness" (see commentary by Frèchet in *The Mathematical Thought*, part IV of Ref. [3]).

[3]I am tempted to assert that such kinds of proofs are actually impossible.

Connection with nature—Nature also presents a sense of irreducibility, but this property acts differently from mathematics. Nature cannot be ultimately reduced to anything different from itself, because it is one whole essence that embeds everything inside itself. But what is for mathematics an intrinsic, defining property, for nature it serves as a delimitation: *mathematics is an expression in the realm of the possible, whereas nature is an expression of what actually is.*

II. INSATURATION: Mathematics Cannot, As a Whole, Be Constructed from a "Master Impredicative"

What it means—First, an "impredicative" is any self-referential definition, and self-reference occurs "when a sentence, idea or formula refers to itself".[4] Second, this conjecture relies on the idea that countably infinite, self-referencing mathematical formulations can always be exhaustively specified,[5] but mathematics itself is not *saturated.* This means that mathematics cannot, as a whole, be constructed from a single, all encompassing self-referential definition, i.e., from a "master impredicative".
Conjecture's nickname—The vertigo extremeness of mathematics.[6]
Connection with nature—Nature also presents a sense of insaturation, but, again, in a delimiting way. How can nature establish itself as a single, all encompassing self-referential essence? Nature is an open system towards becoming, in this sense, it is insaturated against its full self-definition. It opens a kind of "space for realizations to come" [5]: if we attempt to fully characterize nature, all its past must be integrated; yet, the future, which is potentially a part of it, can never enter into its determination, *except as a prediction, as an expectation, and never as a realization.*[7] Is there a "reference" to which the universe is *not* an open system? Any answer to this question must involve a sense of freedom from a delimitation, *but such a freedom cannot be self-referential.*

Further commentaries on the Conjectures—The first conjecture was partially based on observations already raised by the 20th. cent. French philosopher of math-

[4]See, e.g.: http://en.wikipedia.org/wiki/Impredicativity and http://en.wikipedia.org/wiki/Self-reference, respectively.

[5]That statement is assumed valid in terms of *circular quantifications* only. According to Wikipedia, "The vicious circle principle is a principle that was endorsed by many predicativist mathematicians in the early 20th century to prevent contradictions. The principle states that no object or property may be introduced by a definition that depends on that object or property itself." It was later realized that circularity in terms of *quantification* does not lead to paradoxes, as such definitions do not "create sets or properties or objects, but rather just give one way of picking out the already existing entity from the collection of which it is a part" (c.f. http://en.wikipedia.org/wiki/Vicious_circle_principle).

[6]It is just like "looking down the wholeness of oneself from a great height", but, then, with vertigo, one cannot *actually* look.

[7]Here I do not desire to engage in a discussion of various philosophical positions on the nature of time or determinism. Notice that the conjectures do not strictly depend on those.

ematics, Jean Caivaillès.[8] He wrote: "*Mathematics constitutes a becoming, that is, an irreducible reality to something else different from it*" (my translation, from Ref. [3]). Cavaillès' observation, as I interpret it, corroborates with the idea that there is no "meta-language out there", nor "within", which rigorously defines mathematics, *non-mathematically*, to its most basic constituents.[9] Mathematics allows innumerable different representations for the same given concept, relation, or object, and can also be constructed from other basic objects.[10] Yet, any such representation or construction is still completely mathematical. Any precise view of mathematics is mathematical. It all proceeds as if mathematics locked within itself, in order to *effectively realize what it is, as if nothing else could play that unique role.*

The second conjecture was partially based on an informal use of the Gödel's theorems (c.f., e.g., Ref. [4]), as well as on various considerations arising in computer science and mathematical logic. The rationale summarizing those considerations and united in the conjecture is the following. There is a sense in which mathematics is flexible enough to provide a system able to describe an autonomous exhaustion of all consistent logical possibilities within itself. However, it cannot be itself embedded into those same infinitely comprehensive possibilities, that is, *it cannot be constructed from a single and consistent self-definition "on top"*. Such an all-encompassing construction, that is, completely self-referential *and* whole, cannot be "seen" inside the mathematical realm, given whatever consistent set of axioms. Such a consistency can be achieved in principle, but not proven within the system [4]. In a sense, a "view from outside" is required. Hence, mathematics is not saturated, and *specially not so from a "meta point-of-view" of self-referentiality.*

An important note on "autonomous self-referential systems"—I argue that the two conjectures together express limiting conditions for *autonomous self-referential systems*, and that these conditions are optimal, by construction. It is important to note here the reference to "autonomy", in the present context, connected with the notion of self-referential systems. Two observations follow in this note, in order to clarify the principle stated afterwards. First, an autonomous self-referential system is irreducible to anything else that is not itself self-referential, a feature which express a kind of limit: the most elementary self-referential expression must be the primordial one (the generating "seed"), otherwise the system cannot be autonomous, in the sense of self-generating. On the other hand, if the autonomous self-referential system is reducible to a predicative, then it is non-referential, which contradicts the initial assumption that it is. Second, an autonomous self-referential system must contain an assertion "on top" that is not fundamentally self-referential, in order to close

---

[8]Sadly, he had an unfortunate end, being executed by the Gestapo at the end of WWII. Details can be found in Ref. [3], or on Wikipedia, http://en.wikipedia.org/wiki/Jean_Cavailles.

[9]Indeed, it is quite difficult to characterize mathematics, *at the level of rigor that it attains as its natural condition*, by some different means. For example, according to Cavaillès [3]: (i) mathematics is not exactly a part of logic, since some mathematical notions may be combinatorial in nature, or based on other notions irreducible to purely logical operations—yet, those are still mathematical anyway. And (ii) mathematics cannot be entirely characterized as a hypothetical-deductive system, yet alternative/complementary notions for that purpose are still mathematical anyway.

[10]Not necessarily sets [9, 10].

the ladder of all self-referential expressions, thus admitting an *external reference to which it can be defined as autonomous.*[11]

A Principle

The two proposed conjectures reflect "small" and "large" scale features, in an abstract sense, and share a delimiting correspondence with the physical Universe. Indeed, any "reducible essence" of the Universe, whatever its constitution, is always a part of the Universe and cannot be reduced to any "different reality". In the same manner, there is no possible way for the Universe to "become the largest essence of all", under its own unfolding state. The following principle proposes a view for those considerations by stating that mathematics is a *condition*, whereas the Universe is a *manifestation*, and both are inseparable when approached from the point of view of a conscious mind.

PRINCIPLE: Mathematics is an Existence Condition for Autonomous Self-referential Systems, in Particular, The Universe.

What it means—This principle provides a synthesis of both conjectures in the sense that *mathematics uncovers the autonomous self-referentiality of the Universe*, providing, more than a language, a metabolism between conscience and nature.[12] In a stronger sense, *mathematics should arise "with" (note: not "in") any autonomous self-referential universe*, that is, any universe with conditions for the emergence of conscious beings.
Principle's nickname—Mathematics represents the ultimate *tactics* of self-referential systems to mimic themselves.
Explanation—Mathematics does not seem to depend on nature, because there is an indefinitely large number of mathematical constructs and theorems that do not have an immediate application at all, but yet they are eventually discovered and proved independently of the physical world. Many of such mathematical constructions, initially disembodied from a physical application, do end up as being excellent descriptions of new phenomena as soon as we, as humans, construct the necessary relations through observation and theoretical reasoning. Indeed, the human brain is

[11]We allow for the existence of infinitely countable predicatives within the system, as particular cases of impredicatives that are self-referring to a null set.

[12]Far from bringing ideological assumptions to the matter, compare with Marx's quote: "*Labour (...) is an eternal natural necessity which mediates the metabolism between man and nature, and therefore human life itself*" [11]. This made me think of a parallel reasoning for the connection of mathematics and nature.

but an instance of the self-referentiality of the Universe (we are just atoms thinking about atoms [12]). This indicates a firm evidence for the *mathematical knowability of self-referential systems*. This *knowability* is the act of the system to reproduce whatever it is part of, hence, to *mimic itself*.[13] In the present view, mathematics acts as a "conditioner" of all logical possibilites. It is seen as a language only *within the act of making assertions about what is manifested* in the physical world. I believe that this *knowability* indicates that the "reality of mathematics" arises *with* the "reality of the Universe", as a single act. This act is completely crystallized, waiting to be discovered. This *self-intelligibility* is a manifestation of the Universe, *within ourselves*. Mathematics is our tactics to navigate in the ocean that can be known.

Final Reflections and Self-Reflections

The only way for a self-referential system to autonomously exist is to have *bounds* that provide such an autonomy, bounds that I have identified in extreme features of mathematics, delineated in the two conjectures. As stated previously, mathematics is the realm of the possible (condition); nature is the realm of what actually is (manifestation). *The ultimate "tactics" of knowability arises from exhausting that same knowability*. Mathematics and physical laws are, as I attempted to show, a pure self-metabolism, seeking and longing for *that* explanation of itself (*and what shall that explanation be?*).

According to Galileo Galilei,

> Philosophy is written in that great book which ever lies before our eyes—I mean the universe—but we cannot understand it if we do not first learn the language and grasp the symbols, in which it is written. This book is written in the mathematical language, and the symbols are triangles, circles and other geometrical figures, without whose help it is impossible to comprehend a single word of it; without which one wanders in vain through a dark labyrinth.[14]

Indeed, we first learn the language. But, going beyond the language, we turn around to ourselves and see that *we, nature, are the book*, written inside a book, inside another book... A dark labyrinth gives its place to a beautiful spiral. Or a dot. Or a number. Or the infinite. The "effectiveness of mathematics in the natural sciences" is unreasonable and will always be. That is our tactics to keep on turning the pages.

Acknowledgments The author thanks Fabiano L. de Sousa for useful suggestions.

[13]Indeed, one may argue that a sufficiently advanced technology must be indistinguishable from nature. That would be an expected trend, according to the arguments of the present essay. http://www.universetoday.com/93449/do-alien-civilizations-inevitably-go-green/.

[14]http://en.wikiquote.org/wiki/Galileo_Galilei.

References

1. R. Penrose. *The Road to Reality: A Complete Guide to the Laws of the Universe*. Science: Astrophysics. Jonathan Cape, 2004.
2. E. P. Wigner. The unreasonable effectiveness of mathematics in the natural sciences. *Comm. Pure Appl. Math.*, 13:1–14, 1959.
3. J. Cavaillès. *Oeuvres complètes de philosophie des sciences*. Hermann (Edition used here was in Portuguese from Editora Forense, 2012), 1994.
4. Douglas R. Hofstadter. *Godel, Escher, Bach: An Eternal Golden Braid*. Basic Books, Inc., New York, NY, USA, 1979.
5. H. Bergson. *Creative Evolution*. Dover Publications, unabridged edition, 1998.
6. M. Livio. *The Golden Ratio: The Story of PHI, the World's Most Astonishing Number*. Crown Publishing Group, 2008.
7. J. L. Bona, W. G. Pritchard, and L. R. Scott. A comparison of solutions of two model equations for long waves. *NASA STI/Recon Technical Report N*, 83:34238, February 1983.
8. F. Calogero. Why are certain nonlinear pdes both widely applicable and integrable? In JV. Zakharov, editor, *What is Integrability?*, page 1. Springer, New York, 1990.
9. R. Goldblatt. *Topoi, the Categorial Analysis of Logic*. Dover Publications, Inc., 2006.
10. S.M. Lane. *Categories for the Working Mathematician*. Categories for the Working Mathematician. Springer, 1998.
11. K. Marx and L.H. Simon. *Selected Writings*. Classics Series. Hackett, 1994.
12. C. Sagan. *Cosmos*. Ballantine Books, 1985.

Cognitive Science and the Connection Between Physics and Mathematics

Anshu Gupta Mujumdar and Tejinder Singh

Abstract The human mind is endowed with innate primordial perceptions such as space, distance, motion, change, flow of time, matter. The field of cognitive science argues that the abstract concepts of mathematics are not Platonic, but are built in the brain from these primordial perceptions, using what are known as conceptual metaphors. Known cognitive mechanisms give rise to the extremely precise and logical language of mathematics. Thus all of the vastness of mathematics, with its beautiful theorems, is human mathematics. It resides in the mind, and is not 'out there'. Physics is an experimental science in which results of experiments are described in terms of concrete concepts—these concepts are also built from our primordial perceptions. The goal of theoretical physics is to describe the experimentally observed regularity of the physical world in an unambiguous, precise and logical manner. To do so, the brain resorts to the well-defined abstract concepts which the mind has metaphored from our primordial perceptions. Since both the concrete and the abstract are derived from the primordial, the connection between physics and mathematics is not mysterious, but natural. This connection is established in the human brain, where a small subset of the vast human mathematics is cognitively fitted to describe the regularity of the universe. Theoretical physics should be thought of as a branch of mathematics, whose axioms are motivated by observations of the physical world. We use the example of quantum theory to demonstrate the all too human nature of the physics-mathematics connection: it is at times frail, and imperfect. Our resistance to take this imperfection sufficiently seriously (since no known experiment violates quantum theory) shows the fundamental importance of experiments in physics. This is unlike in mathematics, the goal there being to search for logical and elegant relations amongst abstract concepts which the mind creates.

A.G. Mujumdar (✉)
706, Bhaskara, TIFR Housing Complex,
Homi Bhabha Road, Mumbai 400005, India
e-mail: anshusm@gmail.com

T. Singh
Tata Institute of Fundamental Research,
Homi Bhabha Road, Mumbai 400005, India
e-mail: tpsingh@tifr.res.in

A. Aguirre et al. (eds.), *Trick or Truth?*, The Frontiers Collection,
DOI 10.1007/978-3-319-27495-9_18

Physics: Experiments, Unification of Concepts, and Mathematics

Mathematics is a precise language in which true statements can be proved starting from a set of axioms, using logic. (Except for the subtlety posed by Gödel theorems, whose implications for mathematics and physics will not be discussed here). This provides us with the grand edifice of mathematics, with its beautiful and great theorems, and unification across arithmetic, algebra, geometry, and analysis [1]. In mathematics, there does not seem to be a place for experiments. Physics, on the other hand, is an experimental science (hence dependent on technology) of the world we observe, where experiments couple with great leaps of conceptual unification. The mathematics used in physics comes in only at a later stage, when we seek a precise language to describe the observed physical phenomena.

A brief look at the history of physics will testify to these assertions. Thousands of years ago, physics had qualitative, non-mathematical beginnings in primordial human perceptions of matter, light, shape, pattern recognition, space, time, motion, and *elementary counting*. (Elementary counting, or number sense, is hard wired into the brain, as we shall see below, and is a pivotal concept common to both physics and mathematics).

An early observation/experiment of paramount significance had to do with the motion of the sun, the moon, and the planets against the backdrop of the fixed celestial sphere. Elementary knowledge of geometric shapes (such as circles) was used to describe their assumed motion around the earth. Centuries later, it took a great conceptual leap to assert that our description of nature is simpler if the earth and planets are assumed to go around the sun. This was a historic development which was independent of mathematics. Extraordinary and painstaking observations (made possible by advances in technology) established the shape of the orbit of Mars [2]. Then came the concepts of force and acceleration, followed by one of the greatest unifying principles in physics: the force that causes planets to go around the sun is the same as the force of gravity which causes things on earth to fall to the ground. Mathematics came in when Kepler inferred that the Martian orbit is an ellipse. And when Newton deduced from experiments how force and acceleration are quantitatively related. And how the force of gravity falls off with distance, and how the law of motion combined with the law of gravitation shows that planetary orbits around the sun are elliptical. This proof has great mathematical beauty, and the amazing success of such proofs is of course the subject of this discussion. We re-emphasise that in physics, observation/experiment, and concepts and their unification come first, and mathematics makes a later (albeit grand) entry.

Undoubtedly, this character has been repeated throughout the history of physics. Sustained experiments on electricity and magnetism over centuries, and the development of the concept of a field, eventually led to the conceptual understanding that electricity and magnetism are two aspects of the same (electromagnetic) field, to be then followed by the mathematically precise formulation of Maxwell's electrodynamics. The observation that the product of electric and magnetic permeability equals the inverse square of the speed of light led Maxwell to another leap, namely that light is an electromagnetic wave. The fact that classical physics failed to predict the experimentally observed spectrum of black-body radiation suggested to Planck the novel concept that atoms emit and absorb light in discrete quanta—this led to Planck's radiation formula. The inability of classical physics to explain the features of the photo-electric effect suggested to Einstein that light is made of discrete quanta; then came the precise mathematical relation between energy and frequency of the photon. Bohr proposed the concept of quantised angular momentum to explain hydrogen spectra. Finally, Schrödinger and Dirac put the work of Planck, Einstein and Bohr on a firm mathematical footing, in their extremely elegant and universal equations. Always, it is experiments and concepts first, and then the mathematical formulation.

Sometimes, a concept has outlived its time, and must go. The failure of the Michelson-Morley experiment to detect the motion of the earth through the hypothesised ether led Einstein and others to abandon the ether, and look for a set of mathematical coordinate transformations which allow the speed of light to be the same for all inertial observers. Even in the case of general relativity, for which it is said that the theory is built purely on deductive power, one needs to exercise caution while remarking thus. The roots lie in the remarkable experimental fact that objects of all masses fall in a gravitational field with the same acceleration, from which comes the conceptual leap that gravitation is space-time curvature. The leap is followed by years of mathematical struggle on Einstein's part, before he arrives at the correct field equations. Even then, scope is left open for whether or not to include the cosmological constant, and the answer has been provided only now (seemingly so), a hundred years later, by astronomical observations!

On other occasions, one may construct a mathematically consistent theory based on existing experimental knowledge, but the theory is not confirmed by experiments. A case in point is a relativistic scalar theory of gravitation, which is not viable because it fails to predict the bending of light [3]. Or the case of the very elegant and highly symmetric Maxwell equations with magnetic monopoles; to date we do not have experimental evidence for magnetic charges. Hence we do not believe these equations describe nature, even though they are more beautiful than ordinary Maxwell electrodynamics! (Whereas for a mathematician the symmetric equations with monopoles will be 'a thing of beauty is a joy forever'—fun to play with and explore, irrespective of whether or not experiments find monopoles).

The mathematics that physicists use is for the most part relatively simple, only a small subset of the vastness of pure mathematics. Often enough, it is first and second order partial differential equations in many variables, some aspects of algebra and group theory (some groups are more special than others), some aspects of geometry (say Euclidean and Riemannian, as opposed to say affine, projective and Riemann-Cartan geometry), and some aspects of real and complex analysis. Number theory rarely makes an appearance.

One must also note that the mathematics that physics uses is silent about choice of initial conditions! Mathematics describes the laws of physics, which have to be supplemented by initial conditions. What decides the initial conditions of the universe? Are there mathematical laws for them? We do not know, not yet.

We may think of theoretical physics as that subset of pure mathematics whose axioms and concepts are motivated by experiments on the physical world. The consequences of the laws of physics are akin to the theorems that follow from the axioms. Looked at this way, it should not come as a surprise that great physicists sometimes use their equations to predict the outcomes of future experiments. As in Dirac's prediction of anti-matter, and Einstein's prediction of the bending of light. The theoretical law is an all-encompassing mathematically precise description of phenomena that lie in its domain, motivated by experiment, and predictor of experiments to come, until there comes an experiment which exposes the law's limitation, compelling us to look for a more general law.

Mathematics: Primordial Physics, Axioms, Theorems, and Beauty

Remarkably enough, the primordial roots of mathematics are in the same human perceptions as in physics: shape, pattern recognition, counting, space, time, and change. With one significant difference: there is no place in mathematics for matter (material substance), and by extension, for light! This to us is the biggest difference between physics and mathematics, from which all other differences germinate. Matter in mathematics is relevant to the extent that it helps in abstraction, and to arrive at the intuitive notion of a set (of objects). But the kinds of sets more likely to be of interest to a mathematician would be set of integers, set of transcendental numbers, set of all triangles on a plane etc. as opposed to say a physicist's set of planets in the solar system, or set of elementary particles. The commonality of many primordial human perceptions upon which both physics and mathematics build, is to us the reason for the 'unreasonable efficacy of mathematics in physics', as we shall argue in the next section.

Abstracting from counting, shape, pattern, and space-time-change, mathematics receives undefined fundamental entities such as natural numbers, point and line. By giving definitions/axioms, along with operations and relations amongst them, followed by generalisations, mathematicians have developed the classic subjects of number theory, algebra, geometry and analysis. Greater unification, still an ongoing program, is endeavoured through developments such as algebraic geometry, algebraic number theory and arithmetic geometry. This enormous edifice of meaningful and beautiful inter-relations between derivatives of the humble primordial abstractions is, at least on the face of it, very different from physics. Even though physicists use a small part of it to describe laws of the physical world.

Perhaps the most striking example of mathematical abstraction and generalisation, and how it is motivated, comes from the development of the number system. Mathematicians struggled for centuries to understand things which we now teach in high school. Natural numbers were motivated by objects in the physical world, but are quickly abstracted to entities by themselves, with no reference to physical objects. The fundamental laws of arithmetic (commutativity, associativity, distributivity) govern how we add and multiply numbers. The introduction of the zero and subtraction mark important progress. So does the Indian notational system for representing integers, and independently, the principle of mathematical induction. The fascinating and beautiful world of prime numbers, the important proof of the prime number theorem, the still unproven Goldbach conjecture that every even number can be expressed as a sum of two primes, and the still unproven statement that there are infinitely many prime pairs p and $p + 2$. The only recently proven Fermat's last theorem—being the work of many outstanding mathematicians over centuries. Continued fractions, Diophantine equations, and much more [1]. All this makes us believe that numbers have a life of their own. But where do they live? Subsequently we will try to make the somewhat unconventional case that, despite appearances, they live in the human brain, and nowhere else.

Negative integers were introduced in order that subtraction of a larger number from a smaller number became meaningful, while defining operations so that the original laws of arithmetic were preserved. Rational numbers were introduced so that the division b/a is meaningful even if the number b is not a multiple of a. Once again, the generalisation is carried out ensuring the original axioms are preserved by the larger system, else it would be a futile generalisation. The awe-inspiring incommensurables (irrationals) found their place on the number line as non-repeating infinite decimals. The concept of limit, infinite series, and the real continuum were born, and turned out to be an immense benefit to theoretical physics. The arrival of analytical geometry: every geometric object and operation can be mapped to numbers. The mathematical analysis of the infinite, the denumerability of the rationals, the non-denumerability of the continuum, the cardinality of infinite sets, transfinite numbers, the fascinating undecidable status of Cantor's Continuum Hypothesis (that there is no set whose cardinality is greater than that of the set of integers, but smaller than that of the set of real numbers). Cohen's non-Cantorian set theory, where the continuum hypothesis is not true. So far one comes, starting from the primordial perception of counting, having built upon the power of meaningful generalisation. And further afield, the world of complex numbers (still obeying the fundamental laws of arithmetic) necessitated by the need to give meaning to solutions of quadratic equations. The theory of functions of a complex variable flourished; yet the Riemann hypothesis remains the most famous unsolved problem in mathematics, and would have important consequences for our knowledge of prime numbers.

The abstraction of shapes leads to geometry, with no better example than Euclid's historic work in the *Elements*, where the modern mathematical method of deduction from axioms was first laid down more than two millennia ago. With further abstraction, we learn that apart from Euclidean geometry, the plane also admits an affine geometry, which preserves straight lines and parallel lines, but not distances; and projective geometry, which does not preserve parallelism [4]. Abandoning Euclid's fifth postulate, geometers invented curved geometries, paving the way for Riemann's monumental work on hypersurfaces of arbitrary curvature, and differential geometry. Spaces of arbitrary dimension (including infinite) were introduced.

Study of shapes independent of metric and projective properties gave birth to the fascinating field of topology. Euler's formula for polyhedra was an early milestone. The elegant and easy to state Four Color Theorem was proved only as recently as 1977 [5], that too using thousands of hours of computer time, and the proof was later simplified in 1997 and then in 2005. The concept of dimension was generalised as a topological feature. The Hausdorff (fractal) dimension appears intimately in Mandelbrot's theory of fractals. Topological surfaces were classified using concepts such as genus and Euler characteristic. In Falting's theorem, the set of rational points on an algebraic curve make an awesome connection with its genus [6]. The fundamental theorem of algebra can be beautifully proved by considerations of a topological character. The study of equivalence of knots received great impetus from the discovery of the Jones polynomial [7], with deep connections to quantum field theory [8]. In the study of topology, we once again see how abstraction greatly enriches mathematics.

Pattern recognition, with inputs from arithmetic, is the basis of algebra. The fundamental laws of arithmetic are adapted here, giving rise to the concepts of groups, fields, rings and ideals. Each one of them becomes a branch of mathematics by itself. Just to mention one instance, the classification theorem for finite simple groups is more than 10,000 pages long! Algebraic geometry is the modern incarnation of Cartesian analytical geometry: one studies equations for curves and surfaces in higher dimensions, using complex numbers, leading to the definition of algebraic varieties (geometric manifestations of solutions of systems of polynomial equations). The subject has led to some of the deepest mathematical developments and new branches of our times. This is the great twentieth century contribution of Serre and Grothendieck, followed by many others. Arithmetic geometry combines algebraic geometry and number theory. This is a subfield of algebraic number theory which studies algebraic structures related to algebraic numbers. Profound abstraction is now being followed by profound synthesis and unification in pure mathematics.

Calculus was of course abstracted from observation of motion and change, and came with the new concept of the infinitesimal. The origins lay in two more immediate problems: to determine the tangent lines of a curve (the birth of differential calculus) and to determine the area within a curve (the birth of integral calculus). The genius of Newton and Leibniz lay in recognising the connection between the two. It is no wonder that calculus, invented to understand motion, revolutionised mechanics and physics. Mathematical concept and nature showed complete harmony, and today physicists employ calculus, more than perhaps any other branch of mathematics, in their work. The theory of nonstandard analysis borrows sophisticated ideas from modern mathematical logic to make infinitesimals 'respectable' from a mathematical viewpoint.

A unifying feature of mathematics today is that it can be entirely based on an axiomatic treatment of set theory, and derived from a set of axioms, say ZFC (Zermelo-Fraenkel-Choice). But why these particular axioms; especially why the Axiom of Choice? Gödel showed that if ZF is consistent, so is ZFC. Some mathematicians do not like the Axiom of Choice, but most consider that accepting the

Axiom of Choice yields richer mathematics than if one left it out [4]. To us, this in itself is a striking demonstration that mathematics is an enterprise of the human mind, and not a universal Platonic truth ‘out there’.

Based on hard-wired human *primordial perceptions* such as object, size, shape, pattern and change, mankind has built *abstract* concepts such as numbers, point, line, infinitesimal, infinity, equation, group, curvature, and so many more, on which the vastness of mathematics is built. The same primordial concepts give rise to the *concrete* concepts of experimental physics such as force, mass, motion, charge, rotation, and abstract physics/maths concepts such as field and symmetry. (We do not make a distinction between the abstract concepts of theoretical physics and of mathematics). Cognitive science, in its application to mathematics, aims to show in a scientific way how the abstract concepts of mathematics build upon primordial perceptions, using what is known as *conceptual metaphor*. We now argue that when we seek a precise description of inter-relations amongst the concrete concepts on which experimental physics is based, we necessarily rely on the abstract concepts of mathematics. Since the concrete and abstract are both built upon the primordial, this demystifies the extraordinary success of mathematics in physics. It could not have been otherwise.

Cognitive Science, Physics, and Mathematics

In their work, physicists and mathematicians generally prefer to ignore or ‘forget’ the brain, treating it as a perfect and passive agent which helps them discover objective truths about the physical and mathematical world out there. But is that not an unscientific stance? Nature did not evolve the brain for doing physics and maths. How are we ever going to be able to scientifically prove Platonism? Asserting that there is a mathematical universe, which we somehow grasp in an extra-sensory manner, and which perhaps is coincident with the physical universe, is at best an act of faith. Over the last two decades, cognitive neuroscience and cognitive science have made a forceful case that all mathematics is *human* mathematics, made in and by the human brain, and stable across cultures. Whether or not human mathematics coincides with a Platonic mathematics is again not something we can scientifically address. (The coincidence in fact is unlikely, for human mathematics builds on metaphors, whereas Platonism is literal). Physicists use human mathematics to give a precise description of the regularity they observe in their experiments. No surprises here. The great wonder, which lies beyond all present day physics, mathematics and biology, is the very existence of the physical universe with its regularity, and that too a universe with at least one planet full of intelligent beings! How life and the human brain have evolved to the point where such cognitive mechanisms become operational, is by itself a fundamental question, though beyond the scope of the present discussion.

Cognitive neuroscience, which aims to identify the biological substrates which underlie cognition, including mathematical cognition, has made enormous strides in recent years. Fundamental progress in identifying function specificity of brain regions has been made possible by advances in brain mapping techniques such as Positron Emission Tomography (PET) and functional Magnetic Resonance Imaging (fMRI). At the same time, classic laboratory experiments [9] have established that human babies have some innate mathematical abilities. These include, amongst others: ability to discriminate between collections of two and three items (at age three or four days) [10]; one plus one is two (age five months) [11]. These abilities apply to visual as well as auditory stimuli [12]. Similar abilities have been reported in scientific studies on some animals [13–15]. All human beings, independent of culture and education, are capable of subitizing, this being the ability to instantly tell at a glance whether a collection has one, two or three objects [9]. There is some evidence that the region of the brain known as inferior parietal cortex is responsible for symbolic numerical abilities [9]. Remarkably, this is a highly associative area where neural connections from vision, audition and touch also come together. This in itself seems to suggest that primordial physical perceptions (which is based on motor-sensory inputs) might relate to abstractions such as numerics. One can thus say that strong evidence has emerged that some very elementary arithmetic abilities are hard wired into the brain: they are neither discovered nor invented, but they pre-exist in neural connections in infants (The region in the brain known as prefrontal cortex, which is involved in complex motor routines and planning, also seems to be used in complex arithmetic calculations. Rote abilities such as memorization of multiplication tables seem associated with basal ganglia). Cognitive science says that human beings employ ordinary cognitive mechanisms (common with other aspects of understanding and not restricted to mathematics) to go from these simple abilities to advanced mathematics and theoretical physics.

Three key developments in cognitive science which bear crucially on the maths-physics connection are: (i) embodiment of the mind (our concepts are shaped by the nature of our bodies and brains) (ii) role of the cognitive unconscious (our low level thought processes are not amenable to the conscious mind) (iii) role of conceptual metaphors (our abstract concepts are shaped by our primordial perceptions rooted in the motor-sensory system) [16].

Ordinary cognitive mechanisms are shaped by the aforementioned primordial perceptions, which may more formally be classified as: objects and their collections; spatial relations and object distribution in space; time, change and motion; body movements and repeated actions; object manipulation such as rotation and stretching, etc. In technical jargon, there are four relevant mechanisms: (i) image schemas (i.e. spatial relations amongst objects); (ii) aspect schemas (the important discovery that the same neural structure which is used to control complex motor actions is also used in reasoning [17]); (iii) conceptual metaphor: mapping from entities in one conceptual domain to another, say from primordial to abstract, it being a special case of *conflation*, which is the simultaneous activation of two distinct areas of the brain, each of which is concerned with a different experience [18]. Conflation leads to neural connections across domains, resulting in metaphors whereby one domain is conceptualised in terms of another. This is central to understanding how mathematics develops in the brain; (iv) conceptual blends, which combine two distinct cognitive structures [16].

As an example, all of arithmetic, including rationals and negative integers, can be deduced, starting from small numbers, using the following four metaphors: Arithmetic as Object Collection, Arithmetic as Object Construction, The Measuring Stick Metaphor, Arithmetic as Motion along a Path. Algebra is deduced from the metaphor 'Essence (essential properties of a thing) is Form (structure)'. Metaphors have also been proposed for understanding irrational numbers, real and complex numbers, limits and the infinitesimal, transfinite numbers and set theory, and infinity. Path-breaking work has been done in providing a convincing cognitive basis for all of basic mathematics (arithmetic, algebra, geometry, analysis, set theory) [16].

Coming to physics, it is reasonable to distinguish between experimental physics (which describes results of experiments in terms of concrete concepts) and theoretical physics, which uses concrete and abstract concepts to give a mathematical formulation of the observed regularity. Theoretical physics could honorably be called a branch of pure mathematics; that branch whose axioms are motivated by experiments on the physical universe. Concrete concepts include quantitative description of mass, force, space and distance, trajectories, time, rotations, currents etc. Force, for instance, could be metaphorically related to the primordial human perception of the muscular exertion in throwing a stone at a prey or a threat. We have hard wired notions of space and motion, and of the flow of psychological time, which is metaphorically concretised to measured laboratory time. It is thus not difficult to accept that the experiments which physicists perform can be described using concrete concepts which draw upon our primordial perceptions. The abstract concepts used in theoretical physics are either the same as those of mathematics (real and complex numbers, calculus and differential equations, Euclidean and Riemannian geometry) or they are expressly invented using conceptual metaphors. For instance, the abstract concept of a field (gravitational field, electromagnetic field) is metaphored from 'in an event there is a cause and an influence' and the concept 'there can be no action at a distance'. Essentially all of theoretical physics builds on the theme of cause and influence, hard wired into our brains because of the perceived flow of time: (i) how much influence does a source produce at a given location (field equation) and (ii) how does a test object move under this influence (equation of motion). A conceptual metaphor could be: a hunter lights a fire and judges how warm it feels at some distance (field equation) and decides how he should change his position with respect to the fire (equation of motion).

As an example, we argue for the cognitive basis of the laws which determine that the orbit of Mars around the sun is an ellipse: a combination of Newton's second law of motion, and his law of gravitation:

$$m_i \frac{d^2\mathbf{r}}{dt^2} = \mathbf{F} = -G\frac{Mm_g}{r^3}\mathbf{r} \tag{1}$$

Conceptually, it is assumed that Mars, possessed with a heaviness m_i, is in orbit around the sun (having a heaviness M), and its motion is influenced by the sun (which exerts a force diminishing with distance from the sun). No mathematics so far, except in numerical quantification of mass (which is elementary arithmetic, abstracted from innate arithmetic). Only primordial and concrete experimental concepts such as heaviness, motion, force and distance have been introduced. Next, mathematics is introduced by way of the second law, which encodes the experimentally verified inter-relation between the concrete concepts of mass and force, and the abstract entity from calculus (acceleration as the second time derivative of position). The second equality, the force law of gravitation, is motivated, amongst other things, by the necessity to deduce Kepler's empirical inference that the orbit is an ellipse. The equality of the gravitational mass m_g of Mars (that abstract property which causes it to be influenced by the sun) with its heaviness m_i is another great conceptual leap, motivated by terrestrial experiments of Galileo and others. We clearly see in this example how primordial and concrete concepts amongst which a regularity is observed empirically, are mathematically inter-related by the physicist in his/her brain, using additional abstract concepts, to arrive at a law of nature, using mathematics already known or expressly invented for the job at hand.

Similar cognitive structure applies in the rest of theoretical physics. A more advanced example is presented in the next section. The regularity is in the physical world; its mathematical description is a deliberate fit produced in the brain, using known human mathematics, which is based on known cognitive mechanisms. It is tempting but erroneous to conclude that the beautiful mathematical description is resident *in* the physical world out there. Cognition draws upon the physical world to invent the precise, logical and stable human language of mathematics. Human mathematics is in turn used by the mind to give a precise description of the observed regularity of the physical universe. Such a connection is not mysterious. Rather, it is inevitable.

Quantum Theory and Noncommutative Geometry: A Conjecture

Towards the end of the nineteenth century, and in the early twentieth century, experiments such as the spectrum of black-body radiation, the spectral lines of atoms, and the photo-electric effect, severely challenged the classical Newtonian view of mechanics. It quickly became abundantly clear that a new mechanics was needed to explain these phenomena of the atomic and sub-atomic world, and thus quantum mechanics was born. Newton's classical mechanics works for large, classical objects, whereas quantum mechanics works for small objects. Since large classical objects are obviously made up of small quantum objects, we expect classical mechanics to be a limiting case of quantum mechanics. However, even today we do not properly understand the relation between quantum mechanics and classical mechanics [19]. Is this incomplete understanding a hint of another mechanical revolution in the making? Classical $\rightarrow$ Quantum $\rightarrow$ Beyond Quantum? Maybe! Here we employ this tension between classical and quantum to illustrate the cognitive nature of the quantum description of physical phenomena.

One of the central experimental inputs for quantum theory is the relation $E = \hbar\omega$ between energy and frequency, valid both for massless particles (such as photons) and massive particles (such as electrons). This relation makes it mandatory that the quantum state of a quantum particle, say the electron, must be described by a complex number. This is because the energy E of the particle is necessarily positive, and to a traveling quantum wave described by say

$$\psi \sim \exp^{-i\omega t + ikz} \tag{2}$$

one cannot add the complex conjugate, because that will correspond to a negative frequency (and hence negative energy) part, which is disallowed. Thus complex numbers must necessarily enter quantum mechanics if we are to allow for wavelike phenomena, and also allow only for positive frequencies as required by positivity of energy for free particles.

We once again see that first come the experiment on photo-electric effect and the Davisson-Germer experiment establishing the wave nature of massive particles; and then comes the conceptual leap relating energy to frequency. This forces on us the need to describe the state differently, using not real but complex numbers. Cognitive studies trace the origin of complex numbers to the metaphor 'Numbers are points on a line' [16]. Just as a negative integer can be obtained from a positive one by a 180° anticlockwise rotation on the number line, the imaginary number $\sqrt{-1}$ is obtained by a 90° rotation. In this way, an abstract concept such as complex number has been fitted (in the brain) to describe the observed regularity of the quantum energy-frequency relation, and the positivity of energy.

We now dwell on the all too human (as opposed to Platonic) nature of the maths-physics connection in quantum theory. There are four oddities in this connection, bordering on the illogical, which suggest that quantum theory might be incomplete. The first is what we refer to as the dependence on the so-called classical measuring apparatus. We know from text-books that 'when a quantum system interacts with a classical measuring apparatus, the quantum state collapses to one of the eigenstates of the measured observable'. Besides the fact that an entity such as a classical measuring apparatus is vaguely defined, the oddity is that in order to make sense of quantum experiments, the theory has to rely on its own limit, i.e. classical mechanics [20, 21]. This is unsatisfactory: which came first—quantum theory, or its classical limit? The second oddity is that even though the Schrödinger equation is deterministic, the outcomes of a measurement are random, occurring with one or the other probability, given by the Born rule. The initial state is known precisely; there is no initial sampling space, yet the outcomes are probabilistic. This is an extremely unusual state of affairs in physics, the only one of its kind. The mathematical theory of probability is made to stand on its head, so to say! The third oddity is that the collapse of the wave function is instantaneous, so that some kind of acausal influence is felt over spacelike separations, suggesting some tension with special relativity. The fourth oddity is that quantum theory depends on a classical time for describing evolution, which again is unsatisfactory, for how would one formulate quantum theory if there were no background classical space-time?

Despite these oddities, we physicists feel reluctant to modify quantum theory, for the theory is extremely successful, and is not contradicted by any experiment to date. To us, this shows the frailty of the physics—maths connection: even when the link is not fully logical, we are willing to live with it. Once again, experiments reign supreme in physics.

Nonetheless, we venture to ask, in the spirit in which general relativity was developed, how one might improve this connection, motivated on the one hand by ongoing experiments, and on the other hand by the search for a deeper relation with mathematics. Indeed there exists a well-defined phenomenological modification of non-relativistic quantum theory, known as Continuous Spontaneous Localization (CSL) which explains away the first two oddities, and which is being vigorously tested by ongoing laboratory experiments [22, 23].

The need to do away with classical time in quantum theory seems to suggest interesting new links with mathematics. If one cannot have ordinary classical space-time, then there are reasons to believe that it must be replaced by a noncommutative space-time where space-time coordinates do not commute with each other [24]. Such a geometry is a special case of a noncommutative geometry, which is abstracted by first mapping an ordinary geometry to an ordinary commutative algebra, then making the algebra noncommutative, and mapping it back to a new geometry [25].

Doing so opens up two fascinating new frontiers. Firstly, it can be argued that the noncommutative space-time can be described by a noncommutative generalisation of special relativity [26]. Coordinates become matrices/operators, and one has at hand a classical dynamics of matrices, analogous to Adler's Trace Dynamics [27]. The equilibrium statistical thermodynamics of such a theory can be shown to be a quantum theory without classical time, which in an appropriate limit becomes ordinary quantum theory [28]. The consideration of Brownian motion fluctuations about equilibrium seems to lead to the modified quantum theory described by CSL [29]. The oddities of quantum theory seem to go away in a unified manner.

The second remarkable frontier is the connection that noncommutative geometry makes with particle physics and general relativity. The symmetry group G for the Lagrangian of general relativity and the standard model of particle physics is the semi-direct product of the diffeomorphism group and the group of gauge transformations $SU_3 \times SU_2 \times U_1$. Now we ask if there is some space X such that G is the group $Diff(X)$. If so, we will have a geometrization of fundamental interactions. However, a no-go theorem forbids that, so long as X is an ordinary (commutative) manifold. On the other hand, if X is a product $M \times F$ of an ordinary manifold M with a finite noncommutative space F, such a correspondence is possible. The algebra of the space F is then determined by the properties and parameters of the standard model [30, 31].

To us, this ongoing search and endeavour for a better understanding of quantum theory and its relation to space-time structure is one further example of how we humanly abstract from experiments, and then look for appropriate beautiful mathematics to make a better physical theory. It is indeed very hard to envisage that the highly sophisticated mathematics of noncommutative geometry *lives* out there in the extremely abstract space $M \times F$, into which our space-time manifold M has been subsumed. It seems much simpler to accept that the physical laws determined by such a geometry are a product of our mind and live in the human brain.

Outlook: Cognition Versus Platonism

When the known laws of physics are applied to cosmology, they tell us that some fourteen billion years ago, the universe was in a hot, superdense expanding phase consisting of radiation, and matter in the form of elementary particles. What was it like before then? We do not yet know physical laws well enough to predict that. Maybe the universe was created from a singular vacuum, or maybe it has existed forever.

Over time, the expanding fireball cooled, and the elementary particles condensed to first form nuclei, and then neutral atoms. The neutral atoms condensed to form stars, galaxies and planets. Our own planet is about four and a half billion years old; and for reasons that we understand only poorly as of now, a novel phenomenon began on earth about three and a half billion years ago: the origin and evolution of living organisms, possibly from some physically motivated reorganisation of inanimate matter. It took a great deal of evolution, and a long, long time, before human beings emerged, some one hundred thousand years or so ago. Our primordial perceptions were shaped by our need for food, mate, and shelter, and by our need to protect ourselves from other creatures. We started making tools, we started inventing, we started changing our environment. We observed nature, we looked at the night sky, we became curious. Physics and mathematics had begun!

In this scheme of things, we may think of the act of trying to understand the physical universe, and discovering its laws, as a form of interaction of matter with itself: interaction of animate matter (the human brain) with inanimate matter (the dynamic physical world). It seems natural to suppose that what we witness—observations and experiments, and the creation and application of mathematics to explain them through laws—are all a part of the physical universe. It is extremely hard to visualise that there is a mathematical world, separate from the physical world, which the human brain gleans, so as to discover mathematics, and puts this mathematics to use, to formulate physical laws. Where is this mathematical world?

One may now try to put forth various arguments to criticise this cognitive/naturalist standpoint. Firstly, how can we be sure that the human brain can/will know answers to everything? Maybe there is a mathematical world, which the brain is capable of grasping and discovering, but the brain is not (yet) capable of understanding 'where' this mathematics resides. Secondly, in trying to use cognitive science to explain the physics—maths connection, we are trying to use the act of understanding to understand understanding itself! Is it meaningful for the brain to try to understand how it creates ideas—is that not the ultimate in self-reference? Is it not less unreasonable to grant the world of Platonic mathematics its existence (even though we do not know where it exists) so that the brain's problem of self-reference is avoided? Let us take the Platonic repository of mathematics as given, and employ the brain to every now and then go and look in the repository and bring home those facets which we need to do physics. Thirdly, if maths is made in the brain, and is not out there, is its extraordinary richness and stable structure not enormously bewildering? Look at number theory itself. So many patterns, so many theorems, layers upon layers of structure. Is all this made in the brain, and stored there, in the material organisation of neurons? Seems incredible!

Fourthly, maybe it is true that there is no separate Platonic world of mathematics; instead maybe at the most fundamental level, the physical world is identical to the mathematical world. The Dirac equation does not just describe the electron: the Dirac equation *is* the electron. Maybe one day we will discover that the deepest laws of physics are nothing other than theorems in number theory! Is it really true that the mathematical description of material substance is different from the substance itself? And lastly, in a universe where there are no human beings, how could the 'cognitively created' laws continue to hold? Is the motion of Mars around the sun still described by an inverse square law of gravitation?

Before we address these objections, it is important to emphasise that we do not yet have at hand all the knowledge necessary to decisively answer the mystery of the maths-physics connection. We can at best make a strong plausibility case for cognitive science, working on the premise that we should push the boundaries of physics and biology as we know it, before accepting 'extra-sensory perception' as the source of mathematical discovery! That said, we must acknowledge the great distances which still need to be covered in mathematics, physics, cosmology and neurobiology. We often do not appreciate the fundamental importance of unification in mathematics. How are we to say what shape mathematics will take upon the unification of algebra, number theory and geometry? There is a long way to go, before we will have at hand the fundamental laws of physics, if we ever do? The first step being a better understanding of quantum theory, arriving at a quantum theory of gravity, and a theoretical explanation for the properties of elementary particles. Even these well-defined goals seem so far in the future; what then to talk of the new goals which might emerge when these are reached? Will the unified laws of physics be identical with the unified laws of mathematics? Only when we have all the laws of physics and mathematics, can we hope to answer: why is the universe there in the first place; has it existed forever, or if it had a beginning, how did it begin? And we need physicists and biologists to find out how and why life began; how life is defined; how the brain does what it does; what is the physical origin of thoughts and emotions; how does the brain make mathematics; what is the scientific basis of consciousness? Is there life elsewhere in the universe; in particular intelligent life? If there is, then how do they describe the laws of the physical world? Is it the same way that we do (very tempting to believe this) or is it a different way? Wouldn't it be a very egoistic view to think that they understand nature in the same way as we do? This would be a very crucial benchmark in helping us understand if maths is out there, or in the brain. Do we use the same language, same words etc. to express ourselves while conveying our emotions/feelings? There are many ways and everyone picks up a different route. An intelligent life could actually be following the least action principle of describing nature. Are we doing that? Even to achieve that, is there only a single mathematical framework? There are many questions which are unanswered.

Now to the objections against the cognitive standpoint. Could it be that Platonic mathematics exists, and the brain can discover it, but the brain cannot figure out where it is? To us, this is speculation, beyond the realm of the scientific method. Let us come to it when all else has failed, and the cognitive stance shown to not work. Can the brain be used to understand as to how we understand? Is it not simpler to avoid such self-reference, and submit to an external Platonism? This is an extremely subtle matter in our opinion, and the final answer must await significant advances in neuroscience. For now, we would like to appeal to the distinction between the cognitive unconscious (alluded to in Sect. Cognitive Science, Physics, and Mathematics) and the conscious mind. The cognitive unconscious is the source of creativity, where new ideas are born. Things that are understood, become a part of the conscious mind, having first originated in the unconscious in a manner we do not know how to describe. It is entirely possible that when we try to understand the act of understanding, we work at the level of the cognitive unconscious, which being at a level deeper than the mind which relates to sensory perceptions, maybe able to answer as to how we understand. It is a possible idea, to be pursued to its end, before we submit to Platonism. How is creation of mathematics different from creation of a language? We create language to describe nature, day to day events etc.; this does not imply that language is out there. We indeed accept it to be a human creation. In some sense, landscape of poetry/art is even larger than the mathematical landscape, since the former can encompass not only logic but imagination as well. Both describe nature: mathematics in a quantitative way and poetry/art in qualitative form.

Mathematics is extraordinarily rich in structure, and stable, and logical. Can the human brain make all of it, and yet some more? In particular, the vast richness of numbers, their interplay, their inter-relations. Anyone who has played with numbers gets this feeling that they have a life and a world of their own, and that we make contact with this world through our brain, somehow. This feeling that we come away with, is perhaps is the strongest argument in favour of Platonism, and against a cognitive basis for mathematics. Nonetheless, let us pause for a moment to look at the evidence to the contrary, which however little, is growing, and too precious to be ignored. Why do newborn babies and infants have a number sense, which is present much before they start to express themselves, let alone learn to speak a language, and before they are formally introduced to counting, and before the formal process of discovering properties of numbers begins? There are specific regions in the brain which seem to be associated with this number sense, as if to suggest this feature is coded in during the development of the brain in the embryo. Brain lesions are known to specifically impair one or the other numerical ability (multiplication or ordering or something else). Why such precise correlations, unless mathematics is very finely correlated with brain structure and function? Jungle tribes which have not been exposed to formal mathematics also have a feel for numbers. Animals too have a primitive notion of small numbers; that it is much more limited than in humans, and yet not non-existent, suggests number sense is primordially related to the brain. Thus it may not be outlandish to suggest that as we think and build upon the innate number sense, we create universal neuronal patterns which are stable across brains, and which we perceive as 'properties' of numbers. It is perhaps a part of the same grand scheme in the brain that creates universal logic. That logic which creates animate and inanimate matter. Undoubtedly, much more research in neuroscience is needed, but the case for a cognitive basis of number theory is not closed, and must be scientifically pushed till the very end, before we dogmatically accept Platonism.

Perhaps it is worth mentioning here that the very strict logical structure of mathematics applies to deductions from axioms. However, in so far as the axioms themselves are concerned, they come with a flexible human touch. Whether we want to have a particular axiom in a system of axioms is up to us—it is not a matter of logic. One maybe able to do sensible mathematics with an axiom, or without it. Famous examples include Euclid's fifth postulate in geometry; and the Axiom of Choice in set theory. Why should Platonic mathematics permit this freedom in the choice of axioms? In fact, the intuitive, 'primordial' nature of most axioms seems to say much in favour of a cognitive basis for mathematics.

Is the mathematical world the same as the physical world, with the latter being a representation of the former? Is material substance one and the same as its fundamental mathematical description encoded in a physical law? There might be a case for this, if only mathematics were not so infinitely richer than physics. Space is three dimensional, but mathematics allows spaces of any number of dimensions. It seems extraordinarily extravagant to hence suggest that there must actually exist infinitely many different universes with different dimensions. Anything that mathematics allows, must exist in the universe. How are we ever going to test this? Is it not much simpler to suppose that the vastness of mathematics is created by the human brain, and some of it is used to explain the universe? The same way that we dream of so many things in our dreams, but not everything that we dream of is true.

Lastly, in a universe without human beings, what cognitive basis can there be for physical laws? There would be no mathematics in such a universe! Acceptable? We would say yes. Without humans, the universe certainly exists; planets go around the sun, and galaxies drift farther from each other. But there is no concept of the gravitational force, nor a mathematical description of the dynamics. There are no physical laws: we make the laws so as to describe and understand the observed regularity.

There is so much that the human mind creates, and which humans make. People invent tools, gadgets and ideas—result of the finest imagination—wheel, axe, bow and arrow, agriculture, homes, roads, bridges, boats, ships, horse-carriage, paper, language, writing, literature, poetry, art, music, architecture, pen and pencil, guns, tanks, steam engine, train, automobile, aeroplane, spaceship, electricity, lightbulb, telescope, typewriter, computers, photocopiers, watches, smartphones, surgery, medicine … Each and every one of these inventions is rooted in primordial human perceptions, needs, and emotions, and rooted in our interaction with the environment. Why then must we single out mathematics, and ostracise it from this list?! And label the human mind as incapable of inventing mathematics. Much will be gained by trying to understand how the mind makes mathematics, instead of summarily dismissing this as an impossibility. It appears, there is nothing to argue about whether maths is created or invented … the wonder and the deeper question is: could we actually have a universe/multiverse/nature which is permitted through the most generalized form of mathematics and its theorems? And just not limited to what we observe and are capable of discovering through experiments.

References

1. R. Courant and H. Robbins, *What is Mathematics?* (Oxford University Press, 1996).
2. A. Koestler, *The Sleepwalkers: A History of Man's Changing Vision of the Universe* (Penguin Books, 1990).
3. C. W. Misner, K. Thorne, and J. A. Wheeler, Gravitation (W. H. Freeman, 1973).
4. D. Ruelle, *The Mathematician's Brain* (Princeton University Press, 2007).
5. K. Appel and H. Wolfgang, Illinois Journal of Mathematics **21**, 429 (1977).
6. G. Faltings, Inventiones Mathematicae **73**, 349 (1983).
7. V. Jones, Bull. Amer. Math. Soc. (N.S.) **12**, 103 (1985).
8. E. Witten, Commun. Math. Phys. **121**, 351 (1989).
9. S. Dehaene, *The Number Sense: How the Mind Creates Mathematics* (Oxford University Press, 2011).
10. S. E. Antell and D. P. Keating, Child Development **54**, 695 (1983).
11. K. Wynn, Nature **358**, 749 (1992).
12. R. Bijeljac-Babic, J. Bertoncini, and J. Mehler, Developmental Psychology **29**, 711 (1991).
13. F. Mechner and L. Gueverekian, Journal of the Experimental Analysis of Behavior **5**, 463 (1962).
14. R. M. Church and W. H. Meck, *Animal Cognition*, edited by T. G. Bever and H. S. Terrace (Hillsdale, N.J.:Erlbaum, 1984).
15. M. D. Hauser, P. MacNeilage, and M. Ware, Proc. Nat. Aca. Sci. USA **93**, 1514 (1996).
16. G. Lakoff and R. E. Nunez, *Where Mathematics Comes From?* (Basic Books, 2000).
17. S. Narayanan, *Embodiment in language understanding: sensory motor representations for metaphoric reasoning about event descriptions*, Ph.D. thesis, UC Berkeley (1997).
18. C. Johnson, *Metaphor vs. conflation in the acquisition of polysemy: The case of SEE*, edited by M. K. Hiraga, C. Sinha, and S. Wilcox (Amsterdam: John Benjamins, 1997).
19. T. P. Singh, "Youtube video: Does nature play dice?" (2014 https://www.youtube.com/watch?v=wSiDsMKS_uU).
20. J. S. Bell, Physics World, **8**, 33 (1990).
21. L. D. Landau and E. M. Lifshitz, *Quantum Mechanics* (Pergamon Press, 1965).
22. A. Bassi, K. Lochan, S. Satin, T. P. Singh, and H. Ulbricht, Rev. Mod. Phys. **85**, 471 (2013).
23. S. Bera, B. Motwani, T. P. Singh, and H. Ulbricht, Scientific Reports **5**, 7664 (2015).
24. T. P. Singh, Bulg. J. Phys. **33**, 217 (2006).
25. A. Connes, *Noncommutative Geometry* (Academic Press Inc, 1995).
26. K. Lochan and T. P. Singh, Phys. Lett. A **375**, 3747 (2011).
27. S. L. Adler, *Quantum theory as an emergent phenomenon* (Cambridge University Press, Cambridge, 2004) pp. xii+225.
28. K. Lochan, S. Satin, and T. P. Singh, Found. Phys. **42**, 1556 (2012).
29. T. P. Singh, in *The Forgotten Present*, edited by T. Filk and A. von Muller (arXiv:1210.8110) (Springer: Berlin-Heidelberg, 2013).
30. A. Connes, arXiv:math/0011193 (2000).
31. A. H. Chamseddine and A. Connes, Fortsch. Phys. 58, 553 (2010).

A Universe from Universality

Philip Gibbs

A **new paradigm** is emerging in fundamental physics, but what is the true nature of the new thinking? It may be too early to give a full answer but I think that an important part of it is in the way that we understand the vacuum, and how it relates to particle physics. Thirty years ago all physicists would have assumed that the cold flat vacuum is a unique solution of the fundamental laws. The standard model of particle physics is now known to be successful beyond the wildest dreams of physicists back then and it has a single lowest energy state of broken symmetry from which all known laws other than gravity are known to follow. This standard model is thought to be an incomplete theory of particle physics because it does not explain dark matter or inflation. It had been thought that a full unified theory was within our grasp. A fuller unification would be found at the GUT scale from a bigger but simpler gauge symmetry. Beyond that, supersymmetry would provide the final unification with gravity. All the laws of the standard model including its 21 constants would be derived from some unique theory with fewer free parameters, or perhaps even none. Why shouldn't the uniqueness of the vacuum be the way it works in the final solution too?

Today more progressive physicists take a different view. Space and time are seen as emergent from a yet unknown new way of looking at the universe that must go beyond the bounds of standard quantum field theory. Radical as it sounds, that much has been widely accepted for some time and therefore is not the defining feature of the new paradigm. What is harder to accept is the **multiplicity of the vacuum**—the idea that there may be more than one stable solution for cold empty space and that the one we know is nothing special or unique. This concept bruises the egos of particle physicists who thought that the laws of physics they were unveiling were special in a very fundamental sense. They felt that their science is superior in a way that is

P. Gibbs (✉)
Basildon, UK
e-mail: philegibbs@gmail.com

A. Aguirre et al. (eds.), *Trick or Truth?*, The Frontiers Collection,
DOI 10.1007/978-3-319-27495-9_19

different from other fields such as biology or geology. These are understood to be merely studies of one particular solution to the consequences of the laws of physics while many different solutions may be realized elsewhere on other planets. Surely the laws of particle physics could not turn out to be just as parochial.

Problems with the univacuum assumption have been around for a long time. It has been observed for years that the nature of physical laws appears fine-tuned for the convenience of life. With different values for the masses of particles and physical constants, chemistry and astronomy and therefore biology would not be present in the universe. Almost every natural occurring element of the periodic table plays some essential role in the making of multicellular life forms. How can this be explained if there is no alternative possibility for the values of those parameters? This question of anthropic principle was popularised in books by John Barrow and Frank Tipler in the 1980s [1] but physicists had grown used to the form of nature and took it for granted. This changed when they learnt from astronomy that the cosmological constant has a small positive value rather than zero. This is a tuning that is hard to explain using natural arguments. More recently it was the failure to find supersymmetry in the first run of the LHC that rung alarm bells. SUSY is a natural consequence of string theory and would account for **fine-tuning** of the Higgs mechanism, but when the Higgs boson was found in the absence of supersymmetry it was another feature of string theory that seemed to better provide an explanation, namely the multiplicity of the vacuum. When gravity and quantum theory come together we learn that there can be higher dimensions curled up with some undetermined topology and stabilised with fluxes. This provides a **landscape** of different solutions from which the vacuum we know may be selected almost at random. This might not be just a consequence of string theory but may also appear in the so-called alternatives such as spin-foams and Loop Quantum Gravity (I say "so-called" not because I do not accept their viability but rather because I still think they will turn out in the end to be aspects of the same theory from which strings emerge).

There is no shortage of commentators who deny the evidence for fine-tuning. It is true that we only have one example of a universe to study experimentally and predicting the outcome of variations in physical parameters is technically very hard. It is plain that the nature of physical laws would change dramatically if (for example) the strength of the four forces were different, but how do we know that nature would not then self-organise into another universe of equal complexity with new forms of intelligent life? This is not an easy question to answer and many books and papers have been written on the subject that anyone can study. For my part, I am already convinced by what I know, that the accidents of fine-tuning for life that we observe put the case for fine-tuning beyond doubt. If complexity naturally gives rise to life without fine-tuning then we would expect life to be based on a diversity of different natural mechanisms, but on Earth all life forms use the same chemistry of proteins such as DNA. If we find that complex life on other planets follows uniquely the same patterns of biochemistry as ours then the role of anthropomorphism will be undeniable.

Another area of dispute is the success of string theory and its validity as a scientific subject. At the smallest scales we can distinguish two classes of scenario: The spacetime vacuum may be a smooth manifold with its soup of virtual particles, or it may be contorted under the forces of quantum gravity into a complex foam-like quantum structure. We can't tell from low energy experiments which of these two possibilities is right. In the latter case the methods we use in quantum field theory would cease to apply and we would have to find a new way of studying spacetime at the quantum gravity scale just as we needed physical chemistry to describe the structure of matter at the atomic scale. If on the other hand spacetime remains smooth then the consistency arguments we used to build the standard model will continue to apply. These smothness arguments lead to string theory as the only way to describe the gravitons that would be needed to understand quantum gravity in the limit of weak gravity on a flat spacetime. The fact that a solution exists at all in this case under the tight constraints that quantum theory and gravity impose makes a strong case that string theory is correct. We need to take the predicted consequences seriously, even if we don't like what they tells us.

At this stage the multiplicity of vacua remains only a hypothesis, but eventually we will surely understand the theory of quantum gravity well-enough to know if it is correct. In the meantime it is normal and healthy that theorists will build more layers of speculation upon the idea to try and understand the range of possible consequences. It is equally predictable and healthy that such ideas will be criticised. It is all part of the entrance exam that a new paradigm must go through. If there is indeed a class of many possible solutions for the vacuum, is only one of these real? I think it is more parsimonious to accept that **all solutions exist** in some higher sense, whether inside or outside our universe. Some physicists have speculated that there is an eternal process of inflation with vacua decaying to different solutions so that our own universe is just one bubble inside a larger arena. Others have looked at evolving universes where the laws of physics change in leaps where new universes are born from old. We can learn a lot from thinking about such possibilities whether they are eventually testable or not but we should not get carried away by thinking they are less speculative or more testable than they really are.

Some physicists protest that this new way of thinking is "giving up" on fundamental physics. It is thought that string theory allows for as many as 10^{500} vacuum solutions which leads to a belief that it can never predict any experimental outcome. Nothing could be further from the truth. The only thing that is given up is the dream of **naturalness**. This was the idea that there should be unified model of particle physics that explains it all from simple equations based on a unique field theory without any fine-tuning. When Gerard 't Hooft proposed the naturalness idea he was wise enough to not declare it a as a principle. Instead he labelled it a "dogma" with all the connotations that carries.

The figure of 10^{500} different vacua reminds me of a geometric jigsaw marketed by the Vicount Monckton in the UK in 1999. It was claimed that the number of puzzle configurations that would have to be tried to solve it was exactly the same number—10^{500}. A prize of £1 million pounds for a solution was offered with confidence that it would never be claimed, but the prize money was paid out to two ingenious

Cambridge mathematicians who solved it the following year. I think the task of finding the vacuum solution that describes our universe will also be solved, though it may take much longer. It will require new results from experiments such as larger particles colliders, cosmological observatories, gravitational waves and (my favoured hope) proton decay experiments. This could well keep a few more generations of physicists occupied.

At the same time, theorists will desire to look beyond the laws of particle physics that depend on a particular vacuum state and ask "What are the **meta-laws** to which these vacua are solutions?" If the challenges that face particle physics are already hard then this deeper problem may sound like something beyond the possibility of resolution but that seems not to be the case. It is known that the combination of quantum theory and general relativity imposes tough constraints on the possible range of consistent space-time models. Taking perturbations around a flat vacuum we find that the requirement of consistency in quantum field theory limits us to a range of particles with half integer spin up to the spin-two graviton. We can extend to higher spin states using string theory and it is highly likely (but not quite proven) that there is no other way to describe a consistent theory of quantum gravity in perturbation theory. Other approaches such as spin foams and Loop Quantum Gravity approach the quantisation of general relativity from a different direction and tell us more about how to understand space-time in terms that transcend any particular space-time background or particle spectrum.

The ability of theorists to see so far beyond what experiment reveals directly is due to what Wigner called "**the unreasonable effectiveness of mathematics in the natural sciences**" [2] It is not hard to accept that mathematics can describe measurements we make, and tell us what results we will get when we make different but similar measurements, but why is it that mathematics can take us from what we have observed to completely new phenomena not previously looked at? Wigner asked these questions in 1960 citing the use of complex numbers and matrices in quantum mechanics and differential geometry in relativity as examples of the way that ideas from pure mathematics have proved useful in physics. Fifty years later this mysterious power of mathematics is even clearer. Even new ideas from number theory and algebraic geometry have applications in advanced physics.

Critics will say that physicists have been carried away by the beauty of mathematical ideas and used them in ways that are not justified by experiment. This is not the case. Applications of advanced and abstract mathematical ideas appear to arise unexpectedly in physics research and are used as they are needed. Sometimes the physics even moves ahead of what has been previously known in mathematics and advances concepts that mathematicians had been struggling with. Despite this the mysterious nature of the effectiveness of mathematics in physics that Wigner remarked on is not so obvious to people unfamiliar with how mathematics develops. It is claimed by them that mathematics is related to physics simply because the history of the subjects has made it that way. When the Greeks looked at geometry and counting they did so because these were found in nature, but later mathematicians used abstraction to go beyond physics and introduce new ideas like complex numbers, non-euclidean geometry and group theory. You need to get into the mind-set of the pure mathematician to realise the extent to which they have worked independently of any input from nature. This is something that most non-mathematicians are unfamiliar with. Wigner saw that it was deeply mysterious that these things then became useful in physics.

Now we take for granted the use of mathematics such as complex numbers in physics, and the mystery that Wigner saw in them seems to have diminished with time, but new applications of deeper abstract ideas that are less well known to non-mathematicians has replaced them with even greater mystery for those who know the workings. Today we might also equally well ask why physical science is so effective in mathematics. A striking example is the proof by Richard Borcherds of the **Monstrous Moonshine Conjectures** for which he won the Fields Medal. Mathematicians had noticed mysterious and unexpected relationships between numbers and series that arose in unrelated areas of pure mathematics which they called Monstrous Moonshine. Borcherds eventually proved the connection using a clever construction based in string theory. This remarkable connection leaves us to wonder how mathematicians would have proceeded if string theory had not been known from physics. Would they simply have been stuck or would they have invented some purely mathematical form of string theory just to solve this problem? Whatever the answer, it is clear that there are deep and mysterious relations between ideas from physics and mathematics. It is this **unity** that lends hope to the idea that we really can come to understand the meta-laws that govern physics despite the limitations of our technical ability to measure phenomena at the relevant physical scales. Those who deny the mystery need to look deeper to understand that it is very clear to those of us who have immersed ourselves in these subjects.

According to Tegmark's **Mathematical Universe Hypothesis** [3] our universe is just one mathematical structure of many whose existence is equally valid even if they are outside our own universe. It is unnecessary to concern ourselves with whether the words "exist" and "outside" have any meaning here. Such ideas are about concepts beyond our ordinary experience for which we do not have predefined words. To think about them we can only use **metaphors** with meaning that we understand within our own limits. In this sense we can build a picture of this mathematical universe and try to use it to comprehend the nature of reality. Another good way to understand this is to regard mathematical structures as logical possibilities for our universe. Our experience of existence is therefore just one typical realisation of these possibilities from many equally valid alternatives.

In this spirit I offer a **metaphorical chart of the mathematical ontology**. It is a map of all things that are **logically possible** including us and our own universe. It is timeless and pre-geometric because time and space are **emergent** features of our particular universe. In the mathematical ontology there are just relationships between mathematical objects. An ontology is a view of being, of how and why anything exists. I take it as self-evident that logical possibilities "exist" even if only in some metaphorical sense that we don't understand. It is just a way of saying that some things are possible.

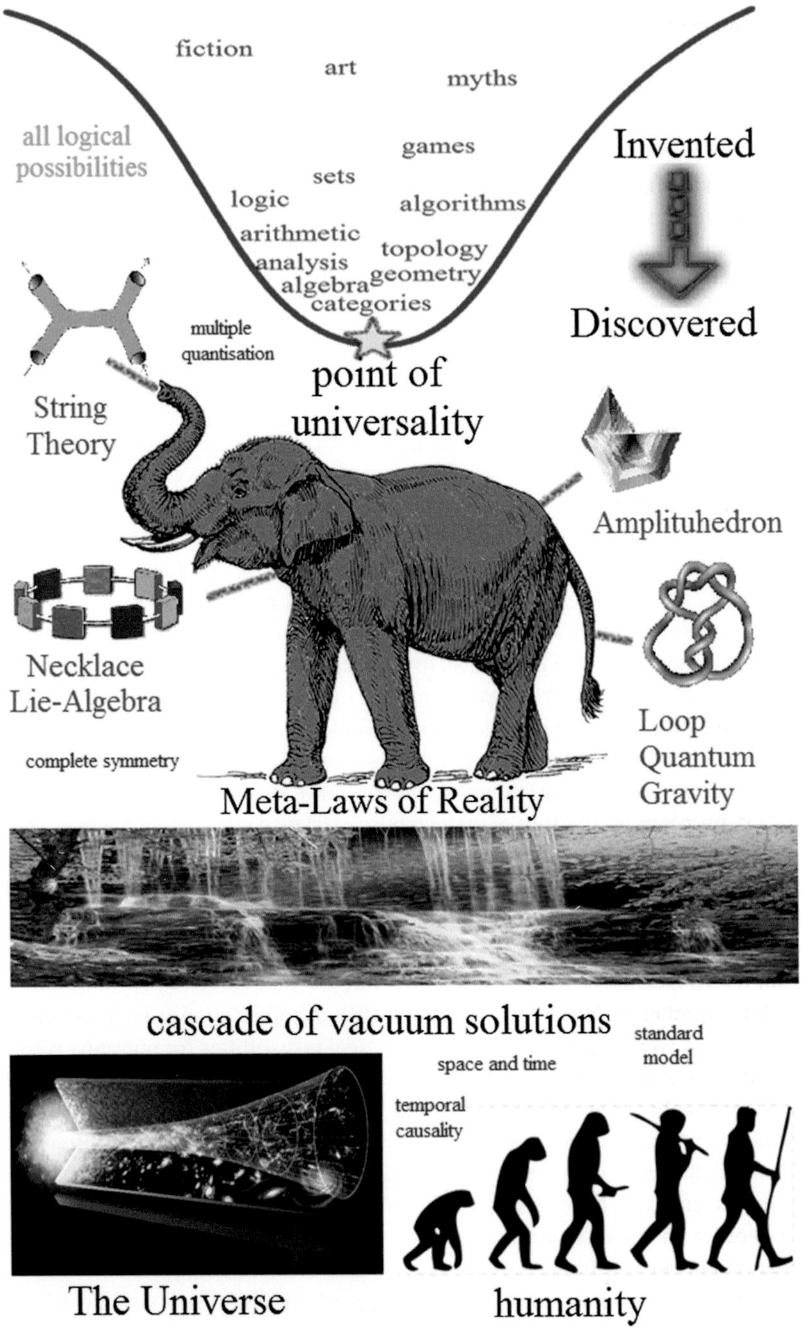

Philosophers sometimes debate whether mathematical structures are **invented or discovered**. This is a key question placed at the top of the metaphorical chart, and the answer is *both on a sliding scale*. A work of fiction has a logical structure both in the language it uses which is represented by a sequence of symbols in a book, and in the relationships and characteristics of the characters and objects in the plot. Mathematicians don't normally regard a story as a work of mathematics because it has a very arbitrary structure that is clearly invented. However, there is no sharp line between such an invention and the more mathematically interesting structures that most mathematicians would describe as discovered. An intermediate structure would be something like the game of chess. It has a very clear mathematical definition and can be analysed mathematically, yet its rules were invented. If we came across an alien civilisation and they played chess we would suspect that there had been some cultural communication that had passed the rules from us to them or vice versa. On the other hand, if the same aliens proved theorems about prime numbers we would assume they probably discovered them independently just as mathematicians often do. What about games with simpler rules like tic-tac-toe or Nim? Somewhere the distinction between invention and discovery blurs and depends on how much time people have had to think of new things to study.

What is it that distinguishes the good mathematical concepts that are discovered from the less interesting ones that seem invented? You might be excused for thinking that the answer lies in simplicity or mathematical beauty and elegance. Those qualities play their part in discovery but they are not the answer. The real answer is something more mysterious. It is referred to as **universality**. If you were a newcomer to mathematics you might expect that the field would consist of a mixture of methods and formula tailored to solve specific problems. If that were the case mathematics would be useful, but it would not be very interesting. What delights mathematicians most is finding connections between problems that had previously seemed unrelated. A mathematical concept like complex numbers or matrices originally formulated to solve one problem in analysis or algebra can turn out to be useful in unexpected areas such as number theory, or of course, in physics. We don't really know why this happens so frequently but it seems to be a feature of universality, something that appears in the study of complexity rather than the study of simplicity. Universality is the quality used to describe mathematical concepts or behaviour that have a wide ranging use or occurrance. It includes things like real numbers, the circle, factorials and the number π. You may think mathematicians are interested in these things because they are derived from some real word structure, but actually those structures just happen to be the first place they were noticed. Other examples of universality such as matrices and complex numbers were first discovered in abstract mathematics.

Even some mathematicians promote the myth that the interesting objects in mathematics are the ones with simple, clear and unique definitions. This is not true, consider for example the number π. The usual definition for π is the ratio of the circumference to the diameter of a circle, but this is not unique. Furthermore, to use this definition you first have to define the geometric concepts such as a circle and the length of a curve. An alternative definition is that π is the first solution to the equation $e^{ix} + 1 = 0$ this still requires some preliminaries but they are easier to formulate. There are many other possible definitions for π so what really makes it a number of interest. The answer is its universality. Its use is ubiquitous in all areas of mathematics, not just geometry. The same is true of the real numbers. It is surprising to people unfamiliar with analysis that defining the real numbers properly starting from axiomatic set theory is a long and complex process. It starts with the natural numbers through the rationals and finally the Dedekind cut. There are many alternatives to Dedekind (Cauchy sequences, Stevin's construction etc.) but they are no simpler and no more obviously the right definition. What makes real numbers interesting is their obvious universal utility.

One of the best examples of a mathematical concept with universality is universal computability. Not all sequences of integers can be computed and the definition for a computable sequence requires a specific language for implementing an algorithm to be selected. Turing used the idea of a Turing machine but that was an arbitrary choice. You could use any computer language to do this and the concept of computability will be equivalent. Some theoretical computer languages might be quite simple with only a few possible instructions, but none is uniquely suited to be singled out as the right definition.

When Carl Sagan wrote his novel "contact" he used the prime numbers as a way that aliens might use to initiate a dialogue with us. This works because the prime numbers area universal concept that any intelligent life would discover and recognise even when he previous contact has taken place. The number π could also be used, perhaps in binary digits. It is reasonable to ask if this common culture of universal concepts depends on us evolving under the same laws of physics. We cant communicate with aliens in another universe but as computer intelligence gets more powerful we will be able to simulate intelligence in a different environment. I predict that a really powerful mathematical intelligence would discover the prime numbers, real numbers, π and all other universal concepts even without the influence of a geometric environment like ours. Mathematical universality is born out of pure mathematical logic and transcends the laws of physics.

Importantly, universality describes emergent behaviour that is found in an unexpectedly wide range of systems. Our understanding of this universality is incomplete, but there are examples of it that give a good general idea. The most revealing is universality found in statistical physics which describes the behaviour of systems consisting of many particles in terms of temperature and entropy which do not depend on the specific microscopic properties of the system. Similar types of universality can be found in quantum field theory where scaling behaviour near a **critical point** in the phase diagram washes out the small scale description of the fields, a feature used in lattice gauge theories to approach a continuum limit.

The mathematical description of statistical mechanics and quantum field theory comes in the form of a weighted sum or integrals over all possible configurations of the system. In quantum field theory we call this a path integral. What then would happen if we treat the whole of mathematics as a statistical physics system or as a **path integral** over the parameter space of all possible theories [4]? Would some universal behaviour emerge that could describe the meta-laws of physics?

That is essentially what I suspect happens on our ontological chart. The realm of all logical possibilities that includes all consistent things whether invented or discovered has a critical point around which universal behaviour can be found. This explains the unity of mathematics and why some concepts seem more discovered than invented, making them of interest to mathematicians. The precise theory of this universality would be algebraic rather than something that can be calculated using statistical physics, but thermodynamic metaphors help us to understand it. As well as explaining the unity of mathematics this universality principle would also explain why mathematics is so effective in natural science. It also explains conversely why physics is so important in mathematicians, even to those who are not interested in practical applications. In my metaphorical chart I show how mathematical concepts funnel towards this critical point of where universality and the meta-laws of physics emerge.

The **meta-laws** themselves are depicted on the chart as an elephant in honour of the ancient metaphor of the elephant and the blind men which is popular in Asian philosophy, especially in India. The moral is that physicists are like blind men (and women) who feel something important as they touch different aspects of the fundamental laws of physics. In reality they are sensing a grand whole which is represented by the elephant. They see string theory when they look at how quantum perturbations of spacetime should propagate as gravitons and they see loop quantum gravity when they apply background independent methods of quantisation to gravity. Other features of the elephant appear as the amplituhedron, non-commutative geometry, spin-foams or twistors. Some of these things are known to be connected but the whole picture still eludes discovery.

Below the elephant we see the **cascade of vacua** which depicts the landscape of solutions to the meta-laws of which our universe is one example. The elephant is not to be envisaged as something that existed before the big bang. Geometry and time are part of specific vacua which can have different dimensions. The emergence of different vacua is ontological rather than temporal [5]. It is possible that different vacua exist in the same universe but a simpler picture is to see them as disconnected possible universes.

Uncovering the meta-laws is now the most important goal in our quest to understand the universe. New empirical data would help but theorists have to work from what they have and the need to combine quantum theory and general relativity under one roof appears to already be a tight constraint. To get answers they must pull together what they know from all the different approaches to quantum gravity and perhaps also from the nature of universality in mathematics.

From string theory there is the idea of **M-theory** from which consistent string theories may be derived as different vacuum solutions. M-theory is thought to be a theory of two and five dimensional membranes in eleven dimensional spacetime. However, there is also F-theory with an extra time dimension and also the bosonic string theories in 26 dimensions. M-theory may therefore be too restrictive to describe the full meta-laws. Instead we should regard M-theory as one aspect of a split in the cascade of vacua, albeit one that is quite high up in the hierarchy.

From work on **Matrix-theory**, the **amplituhedron** and even **Loop Quantum Gravity** we get the idea that space and time are **emergent** so the meta-laws should not be given directly in geometric terms. It has been said that geometry is an angel and algebra is a demon [6], but if so then the signs are that the devil rules at the deepest levels of existence. Perhaps the chart should be turned upside-down. How then can we understand the unholy alliance between these demons and angels that brings into being the graces of space and time from the devilish rules of algebra?

In the 17th century Descartes provided the first components of an answer by defining **coordinates** so that geometric curves could be analysed using algebraic equations. This was followed by vector algebra, matrices, complex numbers, quaternions, tensor analysis and group theory bringing us to the beginning of the twentieth century. Physicists have made use of all those algebraic concepts to understand the geometry of space-time, but during the twentieth century mathematicians such as Grothendiecke moved on to new concepts of **algebraic geometry** that were more abstract and seemed further removed from physics. In the present century physicists have reconnected with the latest mathematical innovations after finding that some of the constructs from algebraic geometry arise naturally in quantum field theories. One of these ideas that I find particularly promising is the application of **iterated integrals** [7] to map algebraic structures like **necklace algebras** [8] to the geometry of particle worldlines, Feynman diagrams, string states and even path integrals. These things are new for both mathematicians and physicists so they will take time to assimilate.

There are people who argue that physicists are misguided and use sophisticated concepts just because of their mathematical beauty which has nothing to do with physical concepts and is not linked to experiment. I think those detractors are wrong and seem to be driven by a desire that the laws of physics should be simple so that more people can understand them. In fact all physicists would be more comfortable if simpler mathematics was all that was needed to find deeper laws of physics because it requires a huge effort to keep learning new things, but what actually happens is that they study quantum field theory and relativity together and find that the hard mathematics just turns up naturally in the analysis. The reason for this confluence of mathematics and physics is mysterious but a possible explanation is that mathematicians and physicists are attracted towards the same critical point of universality. On the mathematicians side it is because the universal constructs are useful across different types of problems which makes them seem more discovered than invented. On the physicists side it is because the point of universality defines the meta-laws of which our own universe is one solution. Of course new experimental input will be needed to confirm that theories built in this way are right but theorists will have to explore the mathematical aspects of quantum field theory combined with relativity in all its forms to know what possibilities can work and which are inconsistent.

What is the role of **symmetry** in the meta-laws? It used to be conventional wisdom that symmetry is the key principle that determines the laws of physics. Group invariance is central to both general relativity and gauge fields so 30 years ago it made sense to look for larger symmetries that unified the forces. Now there is a growing movement among physicists that thinks symmetry is not so fundamental. They say that gauge fields can be regarded as redundant variables that just make the theory look simpler [9]. Different groups control different quantum field theories that are known to be dual to each other so how can the symmetry be what really counts? Many present day physicists prefer to think that symmetry is emergent just as space and time are. To some extent I agree. The fundamental principle that determines the laws of physics is universality, not symmetry, but I contend that symmetry is emergent at the critical point of universality and that it emerges as a **huge symmetry** present in the meta-laws of physics. As we descend the cascade of solutions this universal symmetry is broken by vacuum states and the symmetries we observe at the low energies are residual symmetries, not emergent symmetries. Different dual theories have different symmetries because the universal symmetry is broken in different ways according to the limit taken.

The big clue that this huge hidden symmetry exists is the **holographic principle** that is required to resolve the black hole information loss puzzle. The principle says that the laws of physics can be defined by variables on the surface of a region of space rather than the bulk volume, as if the universe were a giant hologram. This must mean that the when we define the laws of physics in terms of field variable over space-time, those variables must in fact be redundant so that they can be replaced by variables on the boundary. The only mechanism that can realise this is a **complete symmetry** where there is one degree of symmetry for every field variable. In mathematical terms, the physical state is given by the adjoint representation of a symmetry Lie-algebra. Although the universal symmetry is hidden by the selection of the vacuum state the equations of motion for the fields still respect its gauge invariance and can be replaced by an infinite number of gauge charges on the boundary.

Does this mean that the algebraic form of the meta-laws is just a Lie-algebra? The largest and most general Lie-algebra is the **free Lie-algebra** whose universal enveloping algebra is the free associative tensor algebra. It is easy to see that this algebra may play a key role because it takes the form of a necklace algebra (a structure I first encountered in my work on **event-symmetry** [10]) which can be mapped through iterated integrations to both open and closed geometric string states. All other Lie-algebras can be derived from the free Lie-algebra over an infinite dimensional vector space through homomorphic mappings which are the analogue of solution selection in geometry. Could the principles of universal algebra be what governs the meta-laws?

No, it is not quite that simple. There are fermionic fields as well as bosonic fields, so complete symmetry implies a **super-Lie algebra**. Then there are other algebraic generalisation of symmetry that are likely to be relevant. The knotted nature of Loop Quantum Gravity suggests that **quantum deformations** of symmetry such as quantum groups are needed, and from **higher category theory** we know that there are higher dimensional n-groups that are also relevant to mathematical physics. In fact it is the language of these n-categories and operads that has taken over from universal algebra as tools to understand the universal structures of mathematics and there are good indications towards why they also feature in physics. Higher-category theory is the natural description of abstraction. It is at the centre of advanced mathematical research and must be of prime importance in the ultimate understanding of universality. Mathematicians are increasingly finding that n-categories can be used to better understand the structure of M-theory using the higher dimensional n-groups to describe extensions of symmetry [13]. There should be no doubt that this is the right road towards the critical point of universality.

Universality brings together all the logical possibilities of mathematics under one metaphorical **path integral** that quantised the ensemble, but the structure that emerges from universality is also a mathematical structure in its own right. It should also therefore be one of the logical possibilities under the path integral. This gives a **self-referential** and **recursive** nature to universality so we should expect the meta-laws of physics to be not just quantised at one level, but to have the features of multiple layers of quantisation. The relationship between classical and quantum in physics is indeed more complex than it first appears, not only do we have first and second quantisation followed by hints of third and fourth quantisations [11], in addition we find that a given quantum theory can have more than one classical limit. For example theories with S-duality (such as the duality between the electric and magnetic fields in non-abelian gauge theories) have two classical limits with different gauge symmetries related by **geometric Langlands duality**. These are features you would expect if physics is the convergent limit of **iterated quantisations**. When you use Newton-Raphson iteration for calculating square roots you converge to the same final answer for any positive choice of starting value. At each step of the iteration two values map to one value at the next step. We can expect the same to happen with iterated quantisation leading to the same result whether you start from a system of one **qubit** of information or anything else. There are less quantum theories than classical theories and the theory you get by recursively iterating quantisation should be unique.

It is known that fractal (self-similar) structures arise in complex systems. For example, Julia sets appear in chaos theory. The recursive nature of the universe seen through iterated quantisation is a much grander case of the same thing.

The second quantisation process we know from quantum field theory is just a pale shadow of the full algebraic structure present in the meta-laws above the cascade of vacuum solutions. To understand it we need to see quantisation for what it is in its purest form. From one point of view quantisation is the process of taking all mappings from one structure to another that respect their operations, i.e. homomorphisms or n-functors in the more general language of higher category theory [12]. We see this in path-integrals and when we replace classical variables with operators. N-category theory is built in layers of abstraction where n-functors are replaced by $(n+1)$ functors. At the first level of set theory it is set **exponentiation** in which we construct all functions from one set to another. This is why quantisation appears like a process of exponentiation with the exponential function appearing in the path-integral and the number of variables increased as an exponential. Taking this line of reason to its logical conclusion we might find that the meta-laws are described by the **free weak omega-category** for some suitably definition of that object. To understand the laws of physics we need to understand the natural rules of n-category theory and the mappings from algebraic geometry that map the algebraic structures to the geometric ones we are familiar with. It is an ambitious project that will no doubt occupy the minds of physicists and mathematicians for many decades.

The deepest questions we can ask about existence are “How do we exist?” and “Why are things as they are?” The Mathematical Universe Hypotheses tells us that all logical possibilities are equal. It does not require a magic spell to bring one chosen system of equations into reality. Our world is quantum because all those things can happen but it is the principle of universality that makes sense of what we experience. While all is possible in the quantum realm there is a hierarchy of classical limits determined by reversing the self-referential logic of universality. These limits define worlds in which mathematical rules are played out according to the law of quantum averages. They form a landscape of possible solutions of which our universe is just one. Apart from universality the only other constraint is that the solution we experience needs to be such that intelligent life can evolve through **lazy processes** that minimise fine-tuning giving preference to natural structures. Our understanding has been able to progress because universality also determines the elements of mathematics that are most useful and pleasing to mathematicians. The development of both the theory and the experiment is going to be tremendously hard, but during this century we will discover more about the relationships between algebra and geometry that determine the emergence of space and time in a universe governed by the laws of energy and entropy that are needed for life to evolve. At the same time technological progress will enable new empirical observations to help us understand inflation, dark matter, proton decay and other subtle phenomena that help to chart our course through the ontological realm to where we stand in it. They will enable us to pick out the universe’s particular solution to the algebraic meta-laws. Thus we learn finally that there is no mysterious force that defines our consciousness. We have no existence beyond our journey in this material world. This universe of beauty is simply our Heaven or Hell according to the rules of universality and chance and what we make of them in the brevity of our life.

References

1. John D. Barrow, Frank J. Tipler, “The Anthropic Cosmological Principle”, Oxford Paperbacks (1988)
2. Eugene P. Wigner, “The Unreasonable Effectiveness of Mathematics in the Natural Sciences”, Communications on Pure and Applied Mathematics 13, 1, p1–14 (1960)
3. Max Tegmark, “The Mathematical Universe”, Foundations of Mathematics 38, 2, p101–150 (2008)
4. Seth Lloyd, Olaf Dreyer, “The Universal Path Integral”, arXiv:1302.2850 (2013)
5. Philip Gibbs, “The Universe - An Effect Without Cause”, http://fqxi.org/community/forum/topic/1369 (2012)
6. Alain Connes, “A View of Mathematics”, http://www.alainconnes.org/docs/maths.pdf , (2008)
7. Ivan Horozov, “Applications of Iterated Integrals to Number Theory and Algebraic Geometry”, http://www.math.wustl.edu/~horozov/book.pdf (2014)
8. Philip Gibbs “A Universe Programmed with Strings of Qubits”, http://fqxi.org/community/forum/topic/798 (2010)
9. Joseph Polchinski, “Dualities”, arXiv:1412.5704, (2014)
10. Philip Gibbs, “This Time - What a Strange Turn of Events”, http://fqxi.org/community/forum/topic/277 (2008)

11. Philip Gibbs, "An Acataleptic Universe" http://www.fqxi.org/community/forum/topic/1603 (2013)
12. K-G Schlesinger "Towards Quantum Mathematics Part I: From Quantum Set Theory to Universal Quantum Mechanics", Journal of Mathematical Physics, 40, 3, p1344 (1998)
13. S, Hisham, U. Schreiber, "Lie n-algebras of BPS charges". arXiv:1507.08692

And the Math Will Set You Free

Ovidiu Cristinel Stoica

Abstract Can mathematics help us find our way through all the wonders and mysteries of the universe? When physicists describe the laws governing the physical world, mathematics is always involved. Is this due to the fact that the universe is, at least in part, mathematical? Or rather mathematics is merely a tool used by physicists to model phenomena? Is mathematics just a language to tell the story of our universe, a story which could be told with the same or even more effectiveness using another language? Or quite the opposite, the universe is just a mathematical structure?

Introduction

How mathematical is the physical world? Is mathematics essential to describe the universe, or it is just a tool which is not even necessary? There is no unanimously agreed answer to this question. The spectrum of opinions ranges from "mathematics is just a tool helping us to classify our observations" to "the world is nothing but a mathematical structure". Like for any debate, part of the tension between different views is the implicit usage of different definitions. So, to avoid possible confusion, we should define both mathematics and physics.

Is mathematics discovered, or invented? Are mathematical structures entities which have their own existence, eternal and unchanging? Then, how can we know them? If we can access them with our thoughts, shouldn't they then be connected to our minds somehow? Can we discover and explore them? Or it is us who invent them? Or maybe we invent the axioms, and then we discover the consequences? Let's try to find out.

O.C. Stoica (✉)
Department of Theoretical Physics, NIPNE, Bucharest, Romania
e-mail: holotronix@gmail.com

A. Aguirre et al. (eds.), *Trick or Truth?*, The Frontiers Collection,
DOI 10.1007/978-3-319-27495-9_20

A Universe in a Dot

Think at a number. Did that number exist before you picked it, or it is you who invented it? Don't rush with the answer, because it is not as easy as it might seem.

Consider the most brilliant text ever written. Is it the Bible, or the complete works of Shakespeare, or perhaps the complete collection of arXiv articles? Any of these can be written in a computer. The computer uses a code to assign numbers to letters, for example $A \to 65$, $B \to 66$, ..., $Z \to 90$, $a \to 97$, ..., $z \to 122$. Internally, it represents these numbers in binary form, as strings of 0 and 1, so that for example A becomes 01000001, B becomes 01000010 *etc*. Let's take any text and assign to it a number between 0 and 1, of the form $0.d_1d_2\ldots$, where $d_1d_2\ldots$ is the decimal representation of the binary representation of the text.

So, for example, the King James Bible starts with

1. In the beginning God created the heaven and the earth.

Therefore, its number is

$$0.192110078742304518747878005390999191523668554050387237\ldots,$$

which obviously is between 0 and 1 (see Fig. 1).

Not only all literary creations are there, but also all musical works, all movies, anything.

Even this essay is there. So should you stop reading right now, and just look at that line?

That line contains a complete description of your entire life, your past and future, and the moment of your death, since all written information that survives us is nothing but a text but a number but a point.

Imagine now a distant future of mankind, when we know as much as possible about the universe. Suppose we will collect all information we will have about the universe, and also all human creations, recorded in any way, and store everything on a digital support. It's associated number will also contain everything, including any piece of literature, art or science ever written. A universe in a dot (Fig. 2).

By not allowing you to do anything you want, mathematics feels often like a prison. But you have absolute freedom even in a small segment like [0, 1]!

Fig. 1 The Bible as a number between 0 and 1

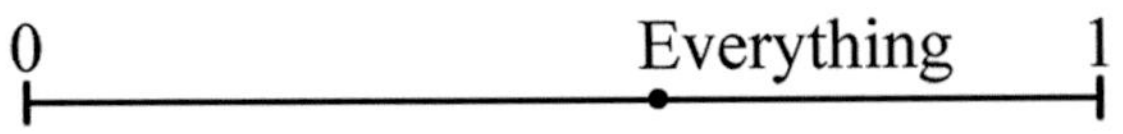

Fig. 2 All human creation and knowledge is a point on this line segment

So, Is Mathematics Discovered, or Invented?

Do the numbers between 0 and 1 already exist, or we keep inventing them? If they preexist, then we have to admit that any poem is already there, among the numbers between 0 and 1. If we invent them, then we have another problem. Given that we label the moments of time with numbers, did we invent every moment from the birth of the universe until now? Since a position in space is given by its coordinates, numbers as well, did we invent every point in space too?

It seems that one cannot simply say that mathematics is invented, since we can't go outside its boundaries, no matter how creative we are. We can't say we discover it either, maybe that we construct it. Or even better, we reconstruct it. It is as if we dance freely, only to find out that we stepped on footprints that were already there. Mathematics is already there, eternal and unchanging. What we invent is the discovery of mathematics.

But What Is Mathematics?

I will take here the position that *mathematics is the science of mathematical structures*. A *mathematical structure* is a set S, together with a collection of relations between the elements of S. A *relation* is a collection of ordered sets of n elements of S [1].

As simple and restrictive as this definition may seem, all the mathematics used by physicists is contained in it. To show how this may be would take many books, but there is room for a few simple examples. The real numbers form a set $\mathbb{R}$. There is a relation of order, which says that 3 is smaller than 4 for example. The relation of order can be seen equivalently as the collection of all pairs of the form (a, b), where $a < b$. How about mathematical operations? Addition is just the collection of triplets (a, b, c), where $a + b = c$. Similarly for multiplication. So, the real numbers form a mathematical structure in the sense defined above. Although I will not detail here how, the entire mathematics currently used in physics is like this, including groups, vector spaces, Hilbert spaces, spacetime manifolds, fiber bundles, and so on.

What Is Physics?

Very roughly speaking, physics should be a collection of true statements about the physical world. It should account for the observations and experimental findings we made so far, but also for those to be made. While we can collect the results of our observations in catalogs, this is not enough to predict the results of future ones.

For example, Tycho Brahe collected the positions of planets and stars from very accurate observations spanned along many years. But to predict the future positions of the planets we had to wait Kepler to discover the rules governing their movement. Unsurprisingly, these were mathematical rules. Due to the work of Newton, the mathematical formulae of Kepler were obtained as consequences of simpler and much more general laws.

Hence, mathematics entered initially in physics as a way to record the observations in a precise way. Then, it was used as an ordering principle, which organizes the quantitative observations. But once they are organized, they appear as consequences of deeper laws which are mathematical in nature. This allows not only to organize the previously known data, but also to make predictions. All physical laws we discovered so far follow the same pattern.

This is why mathematics is so intimately tied with physics.

Is Mathematics Merely a Tool in Physics?

As beings immersed in the universe and observing it, we experience the fact that there are regularities. We learn how to use these regularities to anticipate and control the flow of events. Luckily we don't need them to be truly accurate for most of our daily objectives; it is enough if they are practical.

From this perspective, mathematics seems as a useful tool in organizing our knowledge and giving more or less accurate quantitative predictions. We owe mathematics the freedom brought to us by the scientific and technological progress. But the fact that mathematics is useful doesn't mean the universe is mathematical, not even partially mathematical.

Maybe these useful mathematical models of various phenomena are only approximations of a deeper description. And indeed, when you try to use mathematics to simulate phenomena, you know you use discrete structures to simulate objects that appear to be continuous, you know that the simulation contains a smaller number of degrees of freedom, that it had to ignore some parameters and so on. Yet, the simulation is useful, even though we know it is only an approximation.

But why couldn't that reality we approximate by mathematics be itself mathematical?

Is There Something that Can't be Described by Mathematics?

The idea that mathematics is merely a tool, which is replaceable and is not even the best one, found supporters among philosophers and even physicists. Recently, Smolin and Unger [2–4] brought some strong arguments supporting the idea that mathematics not only is not discovered, but it is not even able to describe the universe in a unified way.

Smolin said [2]:

> there is no mathematical object which is isomorphic to the universe as a whole, and hence no perfect correspondence between nature and mathematics.

> There are no eternal laws; laws are subsidiary to time and to a fundamental activity of causation and may evolve.

Scientists found regularities in the way nature works, regularities successfully tested repeatedly, having predictive power. Mathematics always played a major role in this. We know laws that were never contradicted by our observations, and their validity seems to be eternal, but this doesn't mean they will not be contradicted by experiment someday, or that there are no corners of the universe where they don't apply. So there is no way to disprove Smolin's statements, but there's no way to prove them either. Mathematics instead proved its predictive power together with any successful theory in physics.

Our experience shows that when a law of physics is broken in a particular regime, hence turns out to be approximate and not universal, it is to make room for a more general law, which explains more. For example, the predictions of general relativity went beyond the limits of Newtonian gravity. The laws based on a flat space were replaced with more general laws, on curved spacetime, and the very curvature explained gravity as a form of inertia. When our understanding of general relativity turned out to be limited due to the occurrence of singularities, mathematics provided a more general geometry, one which works at singularities too [5–7], and this allowed us to better understand the big-bang singularity [8, 9], and the black hole singularities [10–13], and to open new avenues to quantum gravity [14].

One central argument invoked by Unger and Smolin is that if mathematical structures would have existence, they would be independent of the physical world. Therefore, any claim that we can describe the universe mathematically would rely on the belief in some mystical powers which would allow us to access worlds outside the universe. The problem with this argument is that people supporting the strong connection between physics and mathematics actually don't think that mathematical and physical worlds are separated, and some claim they are even the same [15]. So, the argument invoked by Smolin and Unger doesn't contradict the hypothesis that the universe is (at least isomorphic to) a mathematical structure.

According to Unger, there are things that can't be described by mathematics, such as *time* and *particularities* [3]:

> Mathematics is an understanding of nature emptying it out of all particularity and temporality: a view of nature without either individual phenomena or time.

But can we find a property of *temporality* that can't possibly be described by mathematics? In fact, time was best understood due to mathematics, in thermodynamics, relativity, and quantum mechanics. Any phenomenon related to time, for example memory, can in principle be described mathematically. Probably what seems to escape any mathematical description is *our feeling* that time flows. But this is another issue, which is not confined to the feeling of time flow only: the *hard problem of consciousness* [16]. However, any feeling we may have, there are neural correlates associated to it, and hence, physical correlates. And these physical correlates are in the domain of known physics, which includes time and is strongly mathematized. But by this I don't claim we can explain consciousness, with or without mathematics. My only claim is that its physical manifestations are describable by mathematics, at least in principle.

Unger writes [3]

> Our mathematical and logical reasoning has a characteristic that places it in sharp contrast to our causal explanations. A cause comes before an effect. Causal explanations make no sense outside time; causal connections can exist only in time.

Actually, there is no conflict between mathematics and causal explanations. The equations describing the time evolution of any physical system reveal the causal connection between the initial conditions and any future state of a system. If causality is still not clear here, consider general relativity, where the very geometry of spacetime gives the causal connections. Of course, the metric tells which events are causally connected, but it doesn't tell the *arrow of time*. The reason why time has a privileged direction is not yet understood, but statistical mechanics explains it by the fact that at a time the entropy was very low. Penrose tried to explain this in a geometric way, and conjectured the *Weyl curvature hypothesis* [17], as accompanying the big-bang singularity. A version of this hypothesis was shown indeed to be true for a large class of big-bang singularities, which are not necessarily homogeneous or isotropic [18].

The other thing which was supposed to be impossible to describe by mathematics are the *particularities*. The reductionist approach aims to explain the diversity of nature in terms of a few fundamental universal laws and building blocks. How is it possible that universal and timeless laws explain particular manifestations which appear, evolve, and vanish? Well, it is not about explaining, unless we take the word "explanation" in its weak sense of "description". Mathematics allows us to describe everything about these particular instances in a reductionist way. But we don't know yet well enough how to obtain from the equations describing how matter evolves in time the various particular forms. Emergent phenomena are still not well understood. But the little we understood so far was done with the help of mathematics, and it doesn't contradict it at all [19–21]. I don't know of a non-mathematical alternative explanation of emergence. At least mathematics can be used to describe the particular forms, even if it can't tell yet how and why they appear. Postulating emergence or particular forms as fundamental is not a solution, we need to explain as much as we can before doing this. Some think even the hard problem of consciousness can be solved in a reductionist way [22, 23], although so far only *easy problems* could be explained. But no matter how scared we may be that mathematics of physical laws is too rigid to allow consciousness, at least we know that there is room for free will, whatever this may be [24–27].

Is Everything Isomorphic to a Mathematical Structure?

From the definition of mathematics I am using here it's easy to see that everything behaves identically to a mathematical structure, everything is isomorphic to a mathematical structure. To see this, consider that we make a complete list of truths about certain domain, which may be even the entire universe. It should be no contradiction between the propositions in the list. Then, there is a mathematical structure for which the same propositions are true. We call the list of propositions a *theory*, and a mathematical structure for which these propositions are true, a *mathematical model* of that theory—in the sense of *model theory* [28].

Some may think that there are things in the universe which can't be described by mathematics. But can you name those things? To name them, you would have to provide a list of their properties, of propositions which hold for them. If the universe is describable by a list of propositions, then there is a mathematical structure describable by the same propositions. But then, couldn't we find something to say which is true about our universe, but not about that mathematical structure? The answer is no. Even if we manage to extend the list with new truths about the universe, there is a mathematical structure which is isomorphic to the universe described by the extended list of propositions [29, 30].

I think at this point many of the readers feel that I am too reductionist, and that there are so many things that can't admit mathematical descriptions. Arts, poetry, love, music, intuition, faith, how can all these be completely mathematical? Well, I don't mean that we have a mathematical description of them. But think at an emotion related to music, poetry, or love. If we describe it completely by a list of true propositions, then you can say it is mathematics. If we can't, it's only because of practical limitations. We know what a feeling is: some chemistry of the brain. But it would be impractical to search for the complete description of the conditions that boost your brain's levels of dopamine, serotonin, oxytocin, and endorphin. However, this doesn't mean that there is no such a description. And if there is, there is also a mathematical model of it, waiting for us.

So, the idea that everything is isomorphic to a mathematical structure may not sound that crazy. Of course, we don't know yet that structure, and it may be too complex to be possibly known, but this doesn't change the things.

If you feel uncomfortable to work with a mathematical structure isomorphic to a phenomenon or a part of the world, you can work safely with the list of propositions describing it, it is the same thing. Sometimes you may prefer to use Tycho Brahe's tables, Kepler's laws, or Newton's laws, depending on the level of abstraction of the problem you want to solve.

Can the Universe be Mathematical?

Even if we admit the idea that our universe is isomorphic to a mathematical structure, does this mean that it is mathematical? Maybe the universe is something that is just describable by a mathematical structure, without being such a structure. But if there is a complete isomorphism between the universe and a mathematical structure (unknown to us, at least so far), there is nothing to tell the difference between them. And since there is no difference that can be tested by experiments, one should admit that the universe is mathematical (I invoke Leinbiz's *identity of indiscernibles* principle). If it looks like a duck, swims like a duck, and quacks like a duck, then it is a duck, isn't it? Whenever we think that there is something more than this, either it has some effects and relations with the rest, which can in principle described by a list of propositions, or it has no effects whatsoever. And if it doesn't have effects, if we limit ourselves to the physical world, we can safely ignore it. So, for now on, when I will say that the universe is mathematical, it will be in this sense.

Does the Hypothesis of a Mathematical Universe Make Predictions?

A *principle* is a general rule which describes in a concise way the data collected in repeated experiments and observations of a class of phenomenona. The principle is a hypothesis, but it can be tested. By using logic, one can derive consequences from the principles, and make predictions of the results of other experiments. If the predictions are not confirmed, we have to reject the principle which led to them.

But is the hypothesis that the universe is mathematical a testable one?

We have seen that any kind of world, as long as it is free of contradictions, is isomorphic to a mathematical structure. This means that this hypothesis is a plain truth that doesn't make predictions at all, and doesn't explain anything. But this doesn't mean it is useless, rather it means that it is the foundation, the framework of any possible theory in physics.

Is There a Theory of Everything?

We know for many decades that our universe is very well described by two theories, general relativity and quantum theory. Each of them has its domain of applicability and validity. But shouldn't be only one theory governing the laws of our universe? It seems natural that this should be true, and that unified theory is often called "the theory of everything", and is supposed to include general relativity and quantum theory and everything else.

Because we couldn't find so far this theory, there are physicists considering that it doesn't exist, and maybe the universe obeys two or even more sets of laws. This doesn't make much sense, since if the universe obeys two or even more independent sets of laws, there must be two or more disconnected mathematical structures modeling them. But we can't live simultaneously in two disconnected worlds. Hence, there must exist a mathematical structure which satisfies our observations about both the quantum world, and the general relativistic one. Maybe these theories are somehow limits of this theory. But the unified theory must exist, even if we don't have it yet.

What Can Stop us from Finding the Theory of Everything?

The standard way of science is this: we look at the data of our experiments and observations, we guess the rule, we derive other consequences of that rule, we design and perform experiments to test those consequences, and if the predictions are invalidated by experiments, we reject the rule and try another one, and so on. But this can't ensure that we will find the ultimate theory of everything (TOE). There is no guarantee that we can test all the truths about the universe. It is clear that, for example, we can't check all the details of events that happened long time ago, but I am not talking here about particular configurations of the universe. What I mean is that it is not sure that we can test even the universal laws. Being able to guess them and then test them would mean either that we are that lucky, or that the universe wants to be completely understood by us, who are just tiny waves on its surface.

However, it may be possible that we will be able to find the fundamental physical laws. The reason is that the universe seems to be very regular. The physical laws appear to be the same at any point and at any time. They appear to be the same for observers moving at different velocities, even though the movement is accelerated. This is the principle of relativity, which is probably the most well tested principle of physics. The laws are the same everywhere, anytime, and for anyone. This simplifies very much the process of testing the other physical laws.

Considering the diversity of phenomena, the number of independent laws which explain almost anything is relatively small. Moreover, each progress that we made unified concepts and replaced the known laws with fewer, more general ones. For example, special relativity unified time and space, energy and momentum, and the electric and magnetic forces. General relativity unified gravity and inertia. Quantum theory unified waves and particles, frequency and energy, explained the atomic spectra. All of the forces in nature, no matter how different they may appear, reduce to gravity (which is spacetime curvature), and three gauge forces. Eventually, all of our knowledge about the physical laws is encoded in quantum physics, the spacetime curvature, and the standard model of particle physics, which involve a much smaller number of concepts than the entire field of physics.

This economy of laws is possible only because of mathematics. When we learn physics, and see how various concepts reduce to more fundamental ones, the general pattern is not only that more complex objects are composed of atoms, which are made of elementary particles. The general pattern is that a handful of laws combine mathematically and give the huge diversity of observed physical phenomena. The best and only way to understand them is with the help of mathematics. Some laws can be explained to non-experts using only words, but with the equations, the things are much clearer, more precise, and avoid naive misunderstandings.

So, if we can understand the universe, it is because this immense complexity can be reduced to a small number of laws. And this reduction is made possible by mathematics.

Some may now say that we will never be able to find the unified theory, because of *Gödel's incompleteness theorem* [31]. The argument is that, since by this theorem any mathematical theory that contains arithmetics is incomplete, this means that we will never be able to find a complete theory of the universe. Well, this is actually a misunderstanding of the famous theorem. Remember that at the end of the XIXth century, it was believed that everything is explained by classical mechanics and electrodynamics. Suppose that there were no relativistic and quantum effects, and the world was really classical. Then, what we had at that time would be the unified theory. Gödel's theorem couldn't prevent this. Did something change after we realized that there are relativistic and quantum effects? Why the Gödel incompleteness theorem would suddenly became relevant? It is true that the unified theory, whichever will be, is likely to contain arithmetic, and hence be incomplete in Gödel's sense, but this would not make it a less unified theory. Gödel's theorem simply states that no finite system of axioms allows any truth valid in that system to be found by finite length proofs.

Hawking gave the following argument against a theory of everything, based on Gödel's theorem [32]:

> we and our models, are both part of the universe we are describing. Thus a physical theory, is self referencing, like in Gödel's theorem. One might therefore expect it to be either inconsistent, or incomplete. The theories we have so far, are both inconsistent, and incomplete.

However, we are just looking for a theory describing the general laws, and not a complete description of this particular instance of the universe, which includes what every human thinks about the universe and themselves. This would not be feasible anyway for practical reasons. A TOE should not be expected to contain the complete description of the state of the universe, only the universal laws, and perhaps various approximate models of particular configurations, for example cosmological models, models of stars and galaxies etc. This is consistent, for the same reason why it is consistent for a computer to contain in its memory the complete technical specifications of itself and its operating system, and an approximate representation of its state. Hence, it is not clear how the fact that we and our theories are part of the world we are trying to describe can lead to the situation in the proof of Gödel's theorem.

Tegmark's Mathematical Universe Hypothesis

The idea that the universe is nothing but a mathematical structure leads to many difficult and interesting questions. For example, why this particular mathematical structure and not another? Tegmark [15, 30, 33] proposes a very democratic point of view, that all possible mathematical structures exist physically too.[1] That is, *mathematical existence* (which means just *logical consistency*—the absence of internal contradictions) equals *physical existence*. Now you may think that to admit that all possible mathematical structures exist is too much of a waste. But the idea is in fact

[1] This is a mathematical expression of David Lewis's *modal realism* [34–36].

very economical, if you think that to choose a single mathematical structure out of an infinity of them requires to specify its definition, or its axioms, while to allow all of them to exist doesn't require to specify all of them, and in fact doesn't require any information. What's easier, to specify the complete works of Shakespeare, the Bible, and any other piece of art, literature, and science, or to specify the interval [0, 1] from Fig. 2?

This makes us wonder where are the other mathematical structures. But the answer is simple: "out there", and if we can't see them, it's because our universe is disconnected from the other possible universes. But if the other structures are disconnected and we can't reach out to check them, then this idea can't be tested.

Tegmark has an interesting idea of a proof of his mathematical universe hypothesis, which in the same time aims to answer the question "why, among all possible worlds, we live in this particular world?". Tegmark's answer is an anthropic one: we live in this world because it is favorable to intelligent observers (*self-aware substructures*, or SAS [33]). For example, the planetary orbits would be unstable if space would have a different number of dimensions [37].

But why the equation of Laplace (which is responsible for the inverse-square law in three dimensions, and hence for the planetary orbits) has to be true, and intelligent beings have to inhabit planets orbiting around stars? Given that we already accepted all possible mathematical structures as physical universes, maybe in the vast majority of the universes the equation of Laplace doesn't hold or is not applicable.

Such arguments that most possible worlds supporting SAS could not be so different than ours rely on varying one constant while keeping the others fixed, and then showing that the optimal value for that constant to allow SAS is the one it has in our universe. But if we vary the entire structure, we can find any kind of worlds which would support one form of intelligence or another.

If we are Turing machines, we could exist in any universe in which such Turing machines can exist. And so many mathematical structures are able to contain simulations of any kind of Turing machines. For example, the *Turing complete* ones can simulate any other Turing machine and any algorithm. Simple cellular automata, governed by very simple rules, like Conway's *game of life* [38] and *Rule 110* [39, 40], are Turing complete. This means that the only prediction made by the mathematical universe hypothesis is that our universe has to be Turing complete, or at least complex enough to simulate intelligences like us. Why would be a universe like ours better at simulating intelligent observers, than the Rule 110 cellular automaton, or than Conway's life game? There is no reason, since they are computationally equivalent. If our intelligence is nothing but an expression of us being Turing machines, we could exist in any cellular automaton which is Turing complete, including in the Rule 110 cellular automaton.

Hence, to prove the mathematical universe hypothesis using Tegmark's idea it is not enough to consider only small variations of our universe, but to take into account many other possibilities to sustain intelligent observers.

Counting Minds

A possibility to modify Tegmark's method of verifying the mathematical universe hypothesis is the following.

Consider the mathematical structures satisfying the following conditions

1. they can be interpreted as Turing machines,
2. they admit configurations which contain self-aware substructures (which in turn can be interpreted as Turing machines).

The notion of *self-aware Turing machines* deserves a rigorous definition, but I don't think we have one at this time. However, they have to be able to gain a significant understanding[2] of themselves and of their universe, and possibly other universes. So I suggest that a pretty good approximation is to consider them to be Turing complete. In this case, any such SAS would be capable of understanding (that is, of simulating) any other SAS. This means that all mathematical structures under consideration are computationally equivalent, so how can we then distinguish them and tell that we live in the one which is most likely to support intelligent observers?

I think the solution is to find a way to count all the possible configurations these mathematical structures can have, and to count among them those containing self-aware observers.[3] For example, among all possible configurations of a certain mathematical structure, find what is the percentage of those containing SAS. Then, given all self-aware substructures, we find in which particular mathematical structures it is more probable for them to exist. If Tegmark is right, the mathematical structures we will find will be very similar to our universe.

The problem is that we don't know how to count the probabilities in the infinite collection of all possible configurations of all possible mathematical structures. For example, if the number of configurations for two structures is infinitely countable, then for any configuration of the first structure there is one for the second structure and the other way around. To find such a correspondence, we label each configuration of the first structure uniquely with the numbers $1, 2, 3, \ldots$, and similarly to the second structure. Then, we put into a one-to-one correspondence the configurations of the two structures having the same label, using the function $f(n) = n$. But we could as well use the function $g(n) = 2n$, and in this case, the second structure may appear to have twice as many configurations as the first structure. To calculate probabilities in infinite sets you need a *measure*, which we don't know how to define in a correct way in this case.

Suppose that we have found indeed that the typical SAS live in mathematical structures similar to our universe. In this case, if another mathematical structure contains SAS, the cheapest way to contain them is by simulating those mathematical structures similar to our universe. But a mathematical structure can exist in two

---

[2]Whatever "understanding" may mean for Turing machines, perhaps "to understand" means "to simulate".

[3]Don't forget to take into consideration the possibility of *Boltzmann brains*—intelligent entities that appeared spontaneously, without a history, because of fluctuations [41].

ways: by itself, or as a substructure of another structure. And to each particular configuration of a mathematical structure which exist by itself, there correspond an infinite ways in which it can exist as a substructure of other structures. For example, if there is one Euclidean plane which exists by itself, there is an uncountably infinite number of Euclidean planes which exist as subspaces of an Euclidean space. And the Euclidean space in its turn is more likely to exist as a subspace of a hyperspace and so on. Therefore, it will be infinitely more likely that SAS exist in mathematical structures which contain simulations of structures like our universe, rather than in structures like our universe themselves.

This only shows that there are important difficulties in trying to prove Tegmark's mathematical universe hypothesis, but not that it is impossible to be proven.

Concluding Remarks

The history of physics shows that as theories evolve, they become more and more mathematized. Mathematizing our theories increases their predictive power, their rigor, their applicability, and unifies various concepts, making them more economical and simpler (but more abstract). Whether this is because mathematics is very versatile, or because physics has a prominent mathematical character, or even because the universe is a mathematical structure, is an open debate.

But I would like to suggest that, just like in our understanding of nature supernatural explanations were gradually replaced by natural ones, in physics, "supermathematical" descriptions are gradually replaced by mathematical ones. I think that we should admit "supermathematical" descriptions as final only if we are sure that we exhausted any hope for a mathematical description. And this may never happen.

Acknowledgments I wish to thank Alma Ionescu for helpful comments.

References

1. G.D. Birkhoff. Universal algebra. In *Comptes Rendus du Premier Congrès Canadien de Mathématiques*, volume 67, pages 310–326. University of Toronto Press, Toronto, 1946.
2. L. Smolin. A naturalist account of the limited, and hence reasonable, effectiveness of mathematics in physics. *Foundational Questions Institute, "Trick or Truth: the Mysterious Connection Between Physics and Mathematics" essay contest, last retrieved September 9, 2015*, 2015. http://fqxi.org/community/forum/topic/2335.
3. R.M. Unger. A mystery demystified: The connection between mathematics and physics. *Foundational Questions Institute, "Trick or Truth: the Mysterious Connection Between Physics and Mathematics" essay contest, last retrieved September 9, 2015*, 2015. http://fqxi.org/community/forum/topic/2332.
4. R.M. Unger and L. Smolin. *The singular universe and the reality of time*. Cambridge University Press, 2014.

5. O.C. Stoica. On singular semi-Riemannian manifolds. *Int. J. Geom. Methods Mod. Phys.*, 0(0):1450041, March 2014.
6. O.C. Stoica. Einstein equation at singularities. *Cent. Eur. J. Phys*, 12:123–131, February 2014.
7. O.C. Stoica. Singular General Relativity. *Ph.D. Thesis*, January 2013. arXiv:math.DG/1301.2231.
8. O.C. Stoica. Big Bang singularity in the Friedmann-Lemaître-Robertson-Walker spacetime. *The International Conference of Differential Geometry and Dynamical Systems*, October 2013. arXiv:gr-qc/1112.4508.
9. O.C. Stoica. Beyond the Friedmann-Lemaître-Robertson-Walker Big Bang singularity. *Commun. Theor. Phys.*, 58(4):613–616, March 2012.
10. O.C. Stoica. Schwarzschild singularity is semi-regularizable. *Eur. Phys. J. Plus*, 127(83):1–8, 2012.
11. O.C. Stoica. Analytic Reissner-Nordström singularity. *Phys. Scr.*, 85(5):055004, 2012.
12. O.C. Stoica. Spacetimes with Singularities. *An. Şt. Univ. Ovidius Constanţa*, 20(2):213–238, July 2012.
13. O.C. Stoica. The Geometry of Black Hole Singularities. *Advances in High Energy Physics*, 2014:14, May 2014. http://www.hindawi.com/journals/ahep/2014/907518/.
14. O.C. Stoica. Metric dimensional reduction at singularities with implications to quantum gravity. *Ann. of Phys.*, 347(C):74–91, 2014.
15. M. Tegmark. The mathematical universe. *Foundations of Physics*, 38(2):101–150, 2008.
16. D.J. Chalmers. Facing up to the problem of consciousness. *Journal of consciousness studies*, 2(3):200–219, 1995.
17. R. Penrose. Singularities and time-asymmetry. In *General relativity: an Einstein centenary survey*, volume 1, pages 581–638, 1979.
18. O.C. Stoica. On the Weyl curvature hypothesis. *Ann. of Phys.*, 338:186–194, November 2013. arXiv:gr-qc/1203.3382.
19. Y. Bar-Yam. A mathematical theory of strong emergence using multiscale variety. *Complexity*, 9(6):15–24, 2004.
20. F. Cucker and S. Smale. On the mathematics of emergence. *Japanese Journal of Mathematics*, 2(1):197–227, 2007.
21. G.F.R. Ellis. Recognising top-down causation. arXiv:1212.2275, 2012.
22. D. Dennett. Are we explaining consciousness yet? *Cognition*, 79(1):221–237, 2001.
23. D. Dennett. *Intuition pumps and other tools for thinking*. WW Norton & Company, 2013.
24. C. Hoefer. Freedom from the inside out. *Royal Institute of Philosophy Supplement*, 50:201–222, 2002.
25. O.C. Stoica. Flowing with a Frozen River. *Foundational Questions Institute, "The Nature of Time" essay contest*, 2008. http://fqxi.org/community/forum/topic/322.
26. O.C. Stoica. Modern Physics, Determinism, and Free-Will. *Noema, Romanian Committee for the History and Philosophy of Science and Technologies of the Romanian Academy*, XI:431–456, 2012. http://www.noema.crifst.ro/doc/2012_5_01.pdf.
27. S. Aaronson. The Ghost in the Quantum Turing Machine. *To appear in "The Once and Future Turing: Computing the World," a collection edited by S. Barry Cooper and Andrew Hodges*, 2013. arXiv:1306.0159.
28. C.C. Chang and H.J. Keisler. *Model theory*, volume 73. North Holland, 1990.
29. O.C. Stoica. The Tao of It and Bit. *Foundational Questions Institute, "It from Bit or Bit from It?" essay contest, fourth prize*, 2013. http://fqxi.org/community/forum/topic/1627.
30. M. Tegmark. *Our Mathematical Universe: My Quest for the Ultimate Nature of Reality*. Knopf Doubleday Publishing Group, 2014.
31. K. Gödel. Über formal unentscheidbare Sätze der Principia Mathematica und verwandter Systeme I. *Monatshefte für Mathematik*, 38(1):173–198, 1931.
32. S. Hawking. Gödel and the end of physics. http://www.hawking.org.uk/index.php/lectures/91, *last retrieved September 9, 2015*, 2002.
33. M. Tegmark. Is "the theory of everything" merely the ultimate ensemble theory? *Annals of Physics*, 270(1):1–51, 1998.

34. D. K. Lewis. *Convention: A philosophical study*. John Wiley & Sons, 2008.
35. D. K. Lewis. *Counterfactuals*. John Wiley & Sons, 2013.
36. D. K. Lewis. *On the plurality of worlds*, volume 322. Cambridge Univ. Press, 1986.
37. M. Tegmark. On the dimensionality of spacetime. *Classical and Quantum Gravity*, 14(4):L69, 1997.
38. J. Conway. The game of life. *Scientific American*, 223(4):4, 1970.
39. M. Cook. Universality in elementary cellular automata. *Complex Systems*, 15(1):1–40, 2004.
40. M. Cook. A concrete view of rule 110 computation. *EPTCS*, 1:31–55, 2009.
41. A. Albrecht and L. Sorbo. Can the universe afford inflation? *Phys. Rev. D*, 70(6):063528, 2004.

Appendix
List of Winners

First Prize
Sylvia Wenmackers: Children of the Cosmos

Second Prizes
Matthew Saul Leifer: Mathematics is Physics
Marc Séguin: My God, It's Full of Clones: Living in a Mathematical Universe

Third Prizes
Most Creative Presentation
Tommaso Bolognesi: Let's consider two spherical chickens
Kevin H Knuth: The Deeper Roles of Mathematics in Physical Laws
Tim Maudlin: How Mathematics Meets the World
Lee Smolin: A naturalist account of the limited, and hence reasonable, effectiveness of mathematics in physics[1]
Cristinel Stoica: And the math will set you free
Ken Wharton: Mathematics: Intuition's Consistency Check
Derek K Wise: How not to factor a miracle.

Fourth Prizes
Alexey Burov, Lev Burov: GENESIS OF A PYTHAGOREAN UNIVERSE
Sophia Magnusdottir: Beyond Math
Noson S. Yanofsky: Why Mathematics Works So Well
Nicolas Fillion: Demystifying the Applicability of Mathematics
David Garfinkle: The Language of Nature
Christine Cordula Dantas: The ultimate tactics of self-referential systems

---

From the Foundational Questions Institute website: http://fqxi.org/community/essay/winners/2015.1

[1] This essay is not included in this volume.

A. Aguirre et al. (eds.), *Trick or Truth?*, The Frontiers Collection,
DOI 10.1007/978-3-319-27495-9

Special Prizes
Non-academic Prize
Philip Gibbs: A Metaphorical Chart of Our Mathematical Ontology
Entertainment Prize
Ian Durham: The Raven and the Writing Desk
Creative Thinking Prize
Anshu Gupta Mujumdar, Tejinder Singh: Cognitive Science and the Connection between Physics and Mathematics
Out-of-the-Box Thinking Prize
Sara Imari Walker: The Descent of Math